ENVIRONMENTAL MICROBIOLOGY

ENVIRONMENTAL MICROBIOLOGY

K. VIJAYA RAMESH

Head
Department of Microbiology
Dr. M.G.R. Janaki College
Chennai 600 028

MJP PUBLISHERS
Chennai 600 005

ISBN 9788180940033 **MJP Publishers**

No. 44, Nallathambi Street,
Triplicane, Chennai 600 005

MJP 002   © Publishers, 2017

Publisher   C. Janarthanan

To
all my teachers

Preface

This book is written for students of life sciences to open their wide doors into the field of applied microbiology where environmental microbiology holds a key position. The book provides fundamental knowledge about the various scopes in applied microbiology like soil microbiology, aeromicrobiology, waste water microbiology and commercial aspects of soil microbiology, such as biofertilisers, biopesticides, etc.

The book is organised into eight units which have easily assimilable chapters that give a comprehensive understanding about various interactions between microorganisms and the environment, be it soil, water or air.

After reading this book, the students will greatly upgrade their fundamental knowledge of the interaction of microbes with the environment and how these interactions are commercialised.

I thankfully acknowledge various esteemed authors of related books from where basic knowledge and new ideas have been gained and presented in this book. I am grateful to the management of my college for providing the basic amenities for completing my work. I am also grateful to Dr. R. Rukumani, Principal, M.G.R. Janaki College, Chennai, for her constant encouragement. I gratefully acknowledge the trouble taken by Mr. Pillai and the staff of MJP Publishers without whose efforts this project would not have been completed. Finally the moral support provided by my family members, colleagues and friends have come a long way with me during this project.

I will appreciate any suggestions from students and teachers to remind me of their needs.

K.Vijaya Ramesh

Contents

PART I

FUNDAMENTALS OF MICROBIAL ECOLOGY

Concepts of Microbial Ecology

Development of Microbial Communities

Succession and Colonisation

1

Concepts of Microbial Ecology

Microbial ecology is the field of science that examines the relationship between microorganisms and their biotic and abiotic environments. Ecology deals with interactions between organisms and relationship between organisms and their environments. Microbial ecology deals only with a segment of the total ecological system. Ecology is a very complex science and is studied from several aspects by the representatives of the various biological sub-disciplines. The use of basic terms and concepts, therefore, is not uniform.

The ecological hierarchy of microorganisms ranges from individuals to an integrated community within an ecosystem. Populations of microorganisms have functional roles (niches) within communities that permit their survival. Microbial populations exhibit various adaptations for success in diverse communities. Communities usually undergo characteristic successional changes that may also lead to greater stability. Disturbances may disrupt the successional process but homeostatic forces act to restore the balance of a community. Stable microbial communities tend to have high diversities. The interactions within the community are often complex; the use of model ecosystems helps to understand population dynamics and ecosystem functioning.

The term *ecology* is derived from the Greek words *Oakes* (house) and *logos* (study). Thus ecology is the science that explores the interrelationships between organisms and their biotic and abiotic environment. In other words, ecology is the study of the interrelationship between organisms in the environments where they live.

Microbial ecology deals with the segment of the total ecological system which specifically examines the relationships between microorganisms and their biotic and abiotic environments.

HISTORICAL ASPECTS

The term *ecology* was first coined by the German biologist Ernst Haeckel (1866). The term *microbial ecology* came into existence only in the early 1960's, but ecologically oriented research on microorganisms was performed as soon as their existence was realised (from the time Leeuwenhoek described microbes).

The work of Pasteur established the importance of microbes in the bio-degradation of organic substances.

Sergie Winogradsky (1890) is regarded as the founder of soil microbiology. He successfully isolated nitrifying bacteria and also developed a model system for growing anaerobic-photosynthetic and microaerophilic soil bacteria now known as the Winogradsky's column. He also developed the concept of *microbial chemoautotrophy*. He described the anaerobic nitrogen fixing bacteria and contributed to the studies of reduction of nitrates and symbiotic nitrogen fixation.He showed that nitrifying bacteria are responsible for transforming ammonia to nitrates in the soil. He originated the nutritional classification of soil microbes as *Autochthonous* (native) and *Allochthonous* (opportunistic).

Martinus Beijerinck (1905) isolated the agents of symbiotic and nonsymbiotic aerobic nitrogen fixers and sulphate reducers. The works of Beijerinck contributed greatly to our understanding of the role of microbial transformations.

The study of Winogradsky and Beijerinck showed that microbe- mediated cyclic reactions are essential for maintaining environmental quality and are necessary for supporting life on earth as we know it.

Beijerinck developed the immensely useful and adaptable technique of enrichment culture.

IMPORTANCE OF
MICROBIAL ECOLOGY

Scientists have recognised that microorganisms occupy key positions in the orderly flow of materials and energy through the global ecosystem by virtue of their metabolic abilities to transform organic and inorganic substances.

Microorganisms have been crucial in solving some of our pressing environmental and economic problems since they have helped us to disperse various liquid and solid wastes in a safe and effective manner.

The most important applications of microbes are in

- recovering metals from low-grade ores (transformation of elements).
- production of food, feed and fuel from by-products and waste materials.
- relieving nitrogen fertiliser shortage.
- biological control of pests.

There are many other practical implications which make microbial ecology a highly relevant and exciting subject for study.

COMMON TERMS USED IN MICROBIAL ECOLOGY

Ecosystem   Basic unit in ecology comprising of abiotic and biotic components.

Biotic element   Community of living organisms dealing with populations of microorganisms.

Population   Consisting of clones of one or several species.

Abiotic   Comprises the chemical and physical conditions in which the microorganisms live.

Habitat   It is the place or locality that a given organism normally occupies (e.g. sediments in lakes, humus rich soils, nasal cavity, intestinal tract of humans, etc.). Most of the microbes have just one habitat within an ecosystem with exceptions (e.g. rhizobia grow in soil and in the root hairs of leguminous plants).

Microhabitat   When the habitat of a species is highly specialised, as seen in certain species of insects (leaf miners) where they live only in the upper photosynthetic layer of the leaves of certain species of plants. Thus the leaf contributes a microhabitat for leaf miners.

Niche   It is the functional role of an organism within an ecosystem. It includes not only where an organism lives but also what it does there. No two population can occupy the same niche at the same time. For example, only those cellulolytic bacteria that can degrade cellulose anaerobically and gain their energy by fermentation can maintain themselves and flourish in the rumen. They must be able to tolerate the temperature of the rumen and the presence of various metabolites.

According to Winogradsky, the microorganisms in an ecosystem can be classified into two categories:

Autochthonous   Those that are indigenous and always present in the given ecosystem (soil, intestine, etc.), and their presence is based on the more or less constant supply of nutrients that are typical for the ecosystem. Their numbers are always constant in any ecosystem.

Zymogenous   Those that are dependent on an occasional increase in concentration of certain nutrients or on the specific presence of certain nutrients. Their numbers in any ecosystem show an increase only when a particular substrate is present in the ecosystem. For example, there is a rise in the numbers of cellulolytic organisms in the soil only when there is a rise in cellulose substrate in the soil. Zymogenous organisms are always found in the ecosystem in very low numbers and show an increase only when there is a proper substrate.

Allochthonous   They are total strangers to any ecosystem. Their presence in any ecosystem is purely transitional and they are not permanent residents of any ecosystem, for example, the spores of soil dwellers present in air for a short period of time (dispersal).

Some examples of ecosystem are pond, lake, root system of plants, oral cavity of man, rumen of cow, segment of intestine, etc.

CONCEPTS OF MICROBIAL ECOLOGY

Energy Flow

Ecosystems are in a dynamic steady state. This is made possible by controlling (by complex feedback mechanism) the numbers or activity of individual organisms, thus maintaining an equilibrium. Alternatively, the population may change in a way that reflects changing levels of matter and energy within the environment.

Energy, usually light, is partly trapped by organisms within the ecosystem and is eventually exported in the form of heat and chemical energy in reduced compounds (carbohydrates). The energy is first trapped by the primary producers which use light in photosynthesis to reduce CO_2. The chemosynthetic bacteria which can use chemicals instead of light energy to produce reduced carbon compounds are also primary producers.

The rate at which energy is stored by primary producers is called the primary productivity and is expressed in energy units (Kcal/m^2/day) or as the amount of carbon or organic matter. Flow of energy through an ecosystem is closely linked with flow of matter.

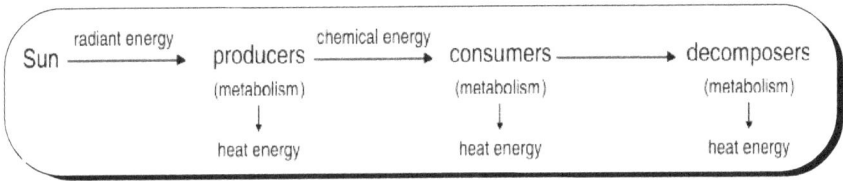

Mechanical energy has two forms:

- kinetic or free energy
- potential energy

Kinetic energy is the energy possessed by a body by virtue of its motion and is measured by the amount of work done in bringing the body to rest.

Potential energy is the stored energy (energy at rest) and becomes useful after conversion into kinetic energy. All organisms require a source of potential energy which is found in the chemical energy of food.

The oxidation of food releases energy which is used to do work. Thus chemical energy is converted to mechanical energy. Hence, food is the means of transfer of both matter and energy.

A portion of the gross primary production is converted back to CO_2 by the respiration of the primary producers. The remaining organic carbon is the net primary production available to heterotrophic consumers who convert the organic carbon back to CO_2 during their respiration. A net gain in organic matter produced by photosynthesis and not converted back to CO_2 is called *net community productivity*.

The transfer of energy stored in organic compounds from one organism to the other establishes a *food chain*. The transfer occurs in steps and each step constitutes a *trophic level*. The interrelationship of food chain steps establish the *food web*.

There are two main ways in which the primary production of an ecosystem may be used by heterotrophic organisms.

Grazing food web   Organisms that feed directly on primary producers (herbivores) constitute the trophic level of grazers. These grazers may be eaten by secondary consumers (carnivores). This is grazing food chain.

Detritus food web   If the primary producers are not eaten alive, then their dead bodies (those of primary and secondary consumers) enters the detritus food chain which is linked copiously to complete the web.

Only 10–15 % of the biomass from each trophic level usually is transferred to the next level. 85–90% is consumed by respiration or enters the detritus food chain.

In most terrestrial and shallow water habitats, the predominant primary producers are higher plants. Microbes do not play important roles as primary producers nor are they generally considered important as consumers. In these habitats 80–90% of total energy flow is through detritus chain. Most of the energy in the grazing food chain is in the bodies of the organism involved.

In the aquatic habitats, the entire food web is based on microbial primary producers, the phytoplanktons (algae, cyanobacteria). Microbes are responsible for most of the ocean's net primary production which is about half of the total photosynthesis of the planet.

The main primary consumers in the aquatic habitat also belong to microbial groups. Most of the primary production (50–90%) goes to the consumers.

The decay portion of the food web (detritus) is dominated by microbial forms in both aquatic and terrestrial environments. The decay portions of the food web involve the degradation of incompletely digested organic matter (faeces, urea, dead plants and animals). Part of the microbial biomass formed during decomposition is recycled into the food web.

Most of the energy in the detritus food chain is stored outside the organism. Hence, this chain is dependent on the primary production in the grazing food chain. The detritus food chain is not solely concerned with decomposition, it contains predators that live on decomposers and this detritus chain is also linked to carnivores of the grazing food chain. Hence, the grazing and detritus food chains are interrelated.

The primary producers and various sorts of heterotrophs in a food chain occupy different trophic levels. In the grazing food chain, microbes occupy the lower levels as primary producers and consumers and animals occupy the higher levels but, microbes decompose the bodies from all levels in the food chain. The higher the trophic level, the less energy it receives, because of respiratory loss and therefore the biomass is usually less than that of lower levels.

In any ecosystem, pyramid of trophic levels based on biomass does not reveal the importance of the organism as much as the pyramid of trophic level based on energy flow through the levels. For example, in a trophic level based on biomass, microbes have no importance since their biomass is exceptionally low but with respect to metabolic activity (energy pyramid) their importance is well ascertained.

Though microorganisms are mostly primary producers or decomposers (saprophytes) in the ecosystem they can also be pathogens since they can obtain nutrients directly from live primary producers, herbivores or carnivores. The trophic levels are not therefore self-contained. The movement between the trophic levels may involve the fundamental changes in metabolism. For example, some algae which are normally photosynthetic primary producers can become heterotrophs when put in the dark with organic matter and then they become part of the decomposer system.

Bio-geochemical Cycle (BGC)

Bio-geochemical cycling describes the movement and conversion of materials by biochemical activities throughout the atmosphere, hydrosphere and lithosphere. A cycle can be described in terms of pools of the substances concerned, so that there is perhaps a pool of soluble material (biotic), a pool of organically bound material and an abiotic atmospheric pool of the gaseous form. BGC cycles include physical transformation (precipitation, fixation, etc.), chemical transformation (biosynthesis, biodegradation, etc.) and various combinations of physical and chemical changes.

All living organisms participate in the bio-geochemical cycling but microbes play a major role in the process. In the BGC cycling, energy is absorbed, converted, temporarily stored and eventually dissipated (energy flows through the ecosystem). This flow of energy is fundamental to the functioning of ecosystem.

When energies flow through the ecosystem, materials undergo cyclic conversions that tend to retain materials within the ecosystem. The cyclic nature of material conversions leads to dynamic equilibrium between various forms of cycled materials.

The BGC cycles usually include negative feedback control mechanisms so that any abnormally high pool level corrects itself and the system returns to the equilibrium state.

Most elements are subjected to some degree of BGC cycling. The intensity or rate of BGC cycling for each element roughly correlates to the amount of the element in the chemical composition of the biomass.

The major elemental composition of living organisms (C, H, O, N, P and S) are cycled more intensely, and trace elements (B, Co, Cr, Cu) and microelements (Mg, Na, K) which are required in small quantities are cycled less intensely.

Microorganisms also act as sources of particular compounds in the ecosphere and sinks for others. Transfer rate between pools vary and are generally enzyme- mediated.

Reservoirs   The various chemical forms of a particular element constitute so called pools or reservoirs and the reservoir sizes can vary greatly from habitat to habitat. Reservoir size is an extremely important parameter to be considered in connection with possible distribution of a cycling system.

In BGC cycling, small, actively cycled reservoirs are the most prone to disturbaces by either natural or human causes.

Attachment Properties
of Microbes

Another important concept of ecology is the adhesion properties of a microbe. All non-toxic, animate and inanimate surfaces have attached microorganisms. The surface need not necessarily be a solid one, even the air-water interface whether it is a bubble surface or the surface of a lake or ocean has attached microbes.

Consequence of phenomenon of adhesion   Nutrients, both organic and inorganic, are adsorbed onto the surface, thus increasing the usual nutrient storage which exists.

Extracellular enzymes may be adsorbed onto the microbial surface. The fact that microbe is attached to something larger may prevent it from being eaten by predators or from being dessicated. Viruses and bacteriophages may survive longer in natural environment if they are adsorbed than when they are free.

Process of attachment   The process of attachment takes place in two stages:

Adsorption   This is physical or chemical and not controlled by the organisms.

Adhesion Organisms secrete a protein or a glycoprotein glue which attracts similar species to each other forming a colony. This colonisation depends on the environmental factors controlling growth

Adsorption is governed by two kinds of forces:

- van der Waal's attraction
- Electrostatic repulsion

Most bacteria and fungi have negatively charged surfaces at neutral pH owing to the anionic groups (COO^-) within their wall polymers. Many soil particles (clay) also bear a negative charge on their surfaces due to isomorphic replacement. Such charged surfaces attract ions of the opposite charge in two layers, one firmly attached and one diffuse layer (100 nm distance from the surface). This distance is greatly reduced (0.5 nm) in concentrated solutions of polyvalent ions. Any surface coating, whether adsorbed or produced by the microorganisms will have a great effect on the surface chemistry.

Hence, at a given ionic strength, the repulsion between the bacterium and a surface bearing a like charge is directly proportional to the reciprocal of the squares of distance between them.

The balance between the van der Waal's forces of attraction and the electrostatic repulsion determines whether the bacterium is attracted to the surface or repelled from it.

Review Questions

1. Define ecology.
2. What is microbial ecology?
3. Brief out the contributions of Sergie Winogradsky and Martinus Beijerinck.
4. Define the following:

    a. Ecosystem   f. Niche

    b. Biotic   g. Autochthonous

    c. Abiotic   h. Zymogenous

    d. Habitat   i. Allochthonous

    e. Microhabitat   j. Reservoir

5. Briefly describe the energy flow in an ecosystem.
6. Brief out on attachment properties of microbes and its importance in ecosystem.

2

Development of Microbial Communities

Community is the *highest biological unit* in a biological hierarchy made up of individuals and populations. A microbial community is an integrated assemblage of microbial populations occurring and interacting with a given location called a *habitat*. Study of community is called *synecology*. Study of the individual population is called *autecology*. Communities vary greatly in the number of *species* they contain (species—a group of organisms with a common gene pool).

DIVERSITY

The heterogenecity of an ecosystem is the diversity. The variety of organisms occurring together in a biological community signifies diversity. Communities are usually characterised by a high state of diversity.

Diversity is related to the complexity of the food web of a community. Food energy can move through a community by a multitude of pathways if the community is diverse. If it is simple, the loss of one or few key species could cause profound changes or even collapse. Similarly, the abnormal rise in the key species may cause an immediate rise in the population of its predator, particularly if it has only one predator that is very narrow in its food requirements.

Biological communities usually contain a few species with many individuals and many species with few individuals. Although a few dominant species normally account for most of the energy flow within a trophic level, the less abundant species determine, in large part, the species diversity of that trophic level and of the whole community. Diversity generally decreases when one or few populations attain high numbers. High numbers signify successful competition and dominance by a single population.

What Determines Diversity

A community that has a complex structure, rich in information as reflected by high species richness (number of species) needs a lower amount of energy for maintaining such a structure. Even though this low energy requirement means a low rate of primary productivity, a stable diversity level is maintained. But, if a community is dominated only by a single species, more energy is needed to maintain them in a stable condition.

Diversity is related to the abundance of ecological niches. An already complex community offers a great variety of potential ecological niches than a simple one. Diversity is inversely related to isolation. For example, islands tend to be much less diverse than ecologically similar continents. This is partly due to the difficulty that many species have in reaching the islands. Diversity is also inversely related to stress and extreme environments. Species diversity tends to be low in physically controlled ecosystems since adaptations to the physicochemical stress are of the highest priority and leave little room for the evolution of closely balanced and integrated species interaction. In acid bogs, hot springs and authentic habitats which are examples of physically controlled habitats, only a few species capable of resisting such conditions will be present.

Diversity is often higher at the margins of distinctive habitats than at the centres because the margins contain all other ecological niches of the habitats. This is called *edge effect*. Diversity is reduced when any one species of organism becomes dominant within a community so that it is able to remove a disproportionate share of available resources thus preventing the growth of many other species. Diversity is greatly affected by biotic history. An area recently vacated by glaciers will have a low diversity because only few species would have had a chance to enter it and become established. A long established stable area might have higher diversity even if it is a poor habitat.

Species Diversity Indices (SDI)

Several mathematical methods that describe the species richness are called Species Diversity Indices (SDI) which are used to describe the assemblage of species diversity within a community. But SDI cannot be applied to the microbial community due to technical difficulties (quantitatively the data is insufficient).

Review Questions

1. Define the following:

    a. Community

    b. Species diversity

    c. Edge effect

2. Discuss at length about species diversity and its influence on ecosystem.

3

Succession and Colonisation

COLONISATION

It is the establishment of a site of microbial reproduction on a material, animal or person without resulting in tissue damage. If the habitat has not been previously colonised (e.g. gastro-intestinal tract of new born) the process is called *primary succession*. When succession occurs in a habitat with a previous colonisation and succession history, it is called *secondary succession* which is a consequence of some catastrophic events (volcanic eruption, barren land, etc.) that has disrupted and altered the course of succession. The first colonisers of a virgin environment are called *pioneer organisms*.

Pre-emptive Colonisation

When pioneer organisms alter the condition in the habitat in ways that discourage further succession, it is called pre-emptive colonisation which may extend the rule of the pioneer organism, but populations better adapted to the newly colonised habitat (altered habitat) usually replace the pioneers. Gradually secondary invaders are also replaced. Succession ends when a relatively stable community called a *climax community* is achieved. It is difficult to apply the concept of climax community to the microbial community. According to the classical ecological thinking, a climax community represents a state of equilibrium and it rarely occurs due to disturbances which randomly affect the successional processes preventing the community from ever reaching full equilibrium.

SUCCESSION

Communities of organisms do not spring into existence suddenly but develop gradually through a series of stages until they reach a state of maturity. This process of *community development* is called *succession*. The individual populations of a community occupy the niches in the ecosystem. With time, some populations are replaced by other populations that are better adapted to fill the ecological niches. Thus succession is the replacement of one community by another as the condition within the habitat changes. The changes may be brought about by the organism itself (by reduction in nutrient and oxygen level or changes in the pH) or they can be imposed from outside (climatic factors). The changes in population are not sudden. The types of interrelationship among populations in a community as well as adaptations within a population contribute to the ecological stability of the community. Some interrelationships involving microbial population are loose associations where one microbial population can replace the other (rhizosphere) whereas others are tight associations where one microbial population cannot replace another (symbiotic association of Rhizobium).

Outcome of Succession

Development of a *more or less stable community* usually involves succession of populations (an orderly sequential change in the population of the community). Long-term seral change (each stage in a succession is called a *sere*) occur in microbial population. For example, bare rocks have a succession of communities starting with algae or lichens as they are converted to stable soils. At the start of the sere, there is stress in the environment (very low nutrient) and the species diversity is low. As the habitat is colonised and amended, the number of species able to grow increases. The final stage in a sere is the climax where there is a dynamic equilibrium between the organism and the environment and usually has great species diversity. Each ecosystem tries to become stable at some point and they tend to be stable in terms of the species present. Though individuals come and go and the balance between the species vary slightly, the overall list of organisms remain stable. Thus stability is not related to diversity but it is imposed from outside, as for example, communities in hot thermal springs have a low species diversity but are very stable because of the environment.

Succession occurs mostly as a result of the influence of physical and chemical changes originating outside the community. Community succession begins with *colonisation* or invasion of a habitat by microbial population.

Types of Succession

Succession can be of the following types:

Autogenic succession In some successional processes, microorganisms modify the habitat in a way that permits a new population to develop. For example, creation of anaerobic conditions by facultative anaerobes allows the growth of obligate anaerobes.

Allogenic succession This occurs when a habitat is altered by environmental factors (seasonal changes), for example, when a salt marsh develops from a tidal mud rich in detritus. Pioneer organisms will be dominated by salt-loving organisms like some algae which will grow and add organic nutrients to the marsh followed by nitrogen fixing algae. Once primary succession is in place, it may be stable for decades. Ultimately coastal marshes can be converted completely into terrestrial communities.

Substrate succession This can be explained with the example of microbes involved in the development of structured soils from bare rock surfaces. The low level of soluble minerals, absence of organic matter and extremes of temperature and moisture on the rocks makes it almost uninhabited and remain a virgin environment. The pioneer organisms could be only the tolerant algae and cyanobacteria with mucilaginous walls or slime capsule and which can fix nitrogen (e.g. *Porphyrosiphon notarisu*, *Gloeocapsa* spp., *Gloecocystis* spp., *Nostoc muscorum*). Organic matter accumulates from wind-blown dust and oxygen, and a surface is established. The algae and lichens are then established followed by fungi and bacteria and then the protozoa. All these processes lead to the build up of primitive thin soil which has more nutrients (organic). It can set back to zero since it is very unstable, unless there is a diversity of microbes and nutrients to support the growth of higher plants.

The sere can be shortened if organic matter is added from outside. Within the long-term sere (from primitive to mature soil) there are many short-term changes in the microbial populations which reflect the colonisation of newly arrived species of organic matter. This is called *substrate succession.*

In soils which have the organic matter in the form of plant litter and animal bodies (soil arthropods) and which arrive with a resident microflora in them, substrate succession usually starts off with the breaking of dormancy of spores affected by the organic matter.

There is a great increase in the zymogenous population (*Mucor, Rhizopus*). Among the bacteria are *Bacillus* and *Pseudomonas* spp. These microorganisms have very high competitive saprophytic ability. As the easily utilisable substrates are consumed and the pH drops due to immobilisation of cations, population slowly changes to ascomycetes, actinomycetes and protozoans which lyse and degrade the fungal hyphae which was a part of initial invasion of substrate. The final stage in the succession is the invasion of the much depleted substrate by autochthonous organisms especially those degrading lignin.

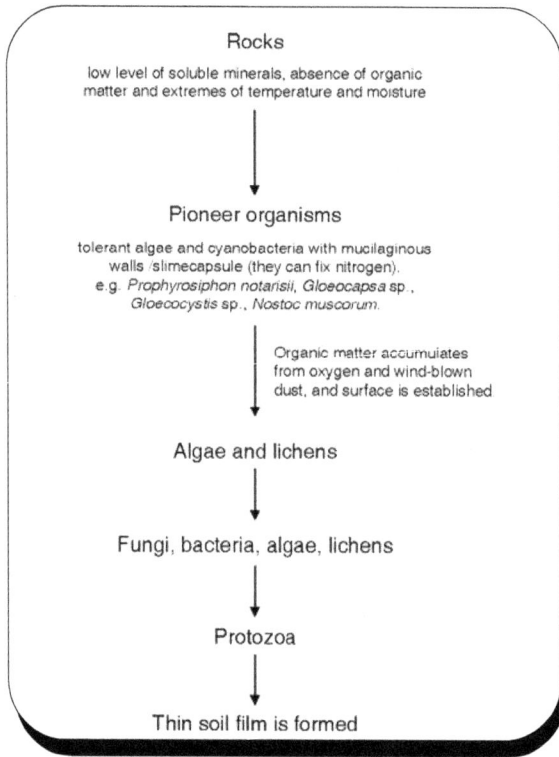

Rocks

low level of soluble minerals, absence of organic matter and extremes of temperature and moisture

Pioneer organisms

tolerant algae and cyanobacteria with mucilaginous walls /slimecapsule (they can fix nitrogen). e.g. *Prophyrosiphon notarisii*, *Gloeocapsa* sp., *Gloeocystis* sp., *Nostoc muscorum*.

Organic matter accumulates from oxygen and wind-blown dust, and surface is established

Algae and lichens

Fungi, bacteria, algae, lichens

Protozoa

Thin soil film is formed

Activity of microorganisms during a substrate succession generally increases the environmental diversity but the substrate and species diversity decreases as easily utilisable nutrients are used up.

Trends in Succession

The various changes in the habitat caused by succession are the following:

- Community productivity goes up

- Biomass tends to increase
- Species diversity of all kinds usually increase for most of the course of succession.

An interesting heterotrophic successional process occurs on detrital particles that enter aquatic habitats. Fresh particulate detritus consists mainly of mechanically shredded tissue of dead leaves, roots, stems, or thalli of macrophytes mixed with smaller amounts of debris from other sources. Microbial communities associated with detritus are complex, but predictable population changes occur during succession. If sterilised natural detrital particles are placed in sea water or fresh water inoculated with a small amount of natural detritus, a characteristic succession of organisms occurs. This succession leads to a microbial community closely resembling that of natural detritus. Bacteria occur in small numbers on the particles after 6–8 hours and reach their maximal numbers after 15-150 hours. The bacterial populations then decrease and become relatively stable after about 200 hours. Small zooflagellates appear about 20 hours after inoculation and reach maximal population sizes after 100–200 hours. Ciliates appear after about 100 hours and reach maximal numbers at 200–300 hours. Other groups of microorganisms including rhizopods and diatoms usually appear late in the succession.

Successions of microbial communities are also associated with animal tissues. The sterile intestinal and skin tissues of newborn animals permit observation of community succession from the time of initial colonisation. The population levels and types of microbes in climax communities in the gastrointestinal ecosystems are regulated by several processes. Some of the regulatory forces in these processes are exerted by the animal hosts, some by the microbes, some by diet, and some by the environment. Within the gastrointestinal tract are many niches filled by various microbial populations. The succession of bacterial populations in humans and other non-ruminant mammals normally begins with colonisation of the gastrointestinal tract by *Bifidobacterium* and *Lactobacillus* species. This is followed by a succession of facultative anaerobes, such as *E.coli* and *Streptococcus faecalis*. Populations of strictly anaerobic bacteria such as *Bacteroides*, appear late in the succession, after the beginning of solid food ingestion. These populations of obligate anaerobes become dominant.

In ruminants, succession leads to the development of a complex, obligately anaerobic microbial community. Included in the climax community of the rumen are populations of cellulose-degrading bacteria such as

Bacteroides and *Ruminococcus*, starch-degrading bacteria such as *Selenomonas*, methanogenic bacteria such as *Methanobacterium*, cellulose and pectin degrading protozoa such as *Polyplastron*, and other populations.

Methanogens are the largest H_2 utilising populations in rumen samples from cattle and in caecal samples from horses. The pioneer bacterial community modifies the environment with the production of various volatile acids and the removal of oxygen, allowing succession to proceed to the climax community.

Review Questions

1. Define the following:

    a. Colonisation

    b. Pre-emptive colonisation

    c. Succession

    d. Sere

    e. Autogenic succession

    f. Allogenic succession

    g. Substrate succession

2. Discuss in detail the outcome of succession and its types, in any ecosystem.

3. What are the changes brought about by succession in any habitat?

PART II

MICROBIAL ECOLOGY OF DIFFERENT ECOSYSTEMS

Structure of an Ecosystem

Freshwater Ecosystem

Marine Ecosystem

Estuarine Ecosystem

Mangrove Ecosystem

4

Structure of an Ecosystem

An ecosystem is the basic functional unit in ecology, as it includes both organisms and their abiotic environment. No organism can exist without the environment. Ecosystem represents the highest level of ecological integration which is energy based. A pond, a lake, a coral reef, part of any field and a laboratory culture can be some of the examples of ecosystems. Thus an ecosystem is defined as a specific unit of all the organisms occupying a given area which interacts with the physical environment producing distinct trophic structure, biotic diversity and material cycling. The term ecosystem was first proposed by the British ecologist A.G. Tansley.

Most of the earth's surface is oceanic, and the deepest part of these oceans is deeper than the highest peaks of the mountains. The marine environment is extensive and complex, containing a vast array of communities and is divided into two major provinces¾estuaries and deep ocean. Oceans are the ultimate sinks for all water soluble minerals, and are saline. Apart from these, mangroves also form part of the marine ecosystem which is a rich source of biotic and abiotic elements.

There are two basic processes in an ecosystem. One of the processes involves a cycle of exchange of materials between living things and the environment. The plants synthesise complex organic materials from the raw materials. The organic matter ultimately releases the raw material which are returned to the environment. This mechanism is called *cycling of materials*. The other basic requirement of an ecosystem is the constant input of energy. The ultimate source of energy is the sun whose solar energy is captured by green plants. Other organisms derive their nutrition and energy from the plants. The energy taken by these organisms is passed on to other organisms.

In this way energy is transferred from one organism to another. This is called *flow of energy*.

The major aspects of an ecosystem are its structure and function. Structure involves:

1.   composition of biological community including species, numbers, biomass, life history and distribution in space, etc.

2.   quantity and distribution of non-living materials, such as nutrients, water, etc.

3.   range, or gradient of conditions of existence, such as temperature, light, etc.

Functions of an ecosystem involves:

1.   rates of biological energy flow, i.e. the production and respiration rates of the community.

2.   rates of materials or nutrient cycles.

3.   biological or ecological regulation including both regulation of organisms by environment and regulation of environment by the organisms. Thus in any ecosystem, structure and function are studied together.

Odum (1959) classified the abiotic components of an ecosystem into three parts:

i.   inorganic nutrients like C, N, H, etc.

ii.   organic compounds constituting the organism.

iii. climatic factors.

The term abiotic means without life or nonliving. Many substances such as water, oxygen, sodium chloride, nitrogen and carbon dioxide are abiotic when they are physically outside living organisms, such as in air or water, but once within living organisms they become part of the biotic world. Many elements may be tightly bound in inorganic compounds as silicon in sandstone or aluminum in felspar, and are unavailable to living organisms. Elements such as oxygen which are normally very active in biological processes may be in an abiotic form readily available to living organisms such as free O_2, CO_2, etc., or they may be in an inaccessible form as silicon dioxide in quartz, a major component of granite. One of the most important aspects of an ecosystem is the rate of release of nutrients from solids, as this regulates the rate of function of the entire system.

The abiotic components can be classified into three groups:

1. *Climatic regime*¾includes temperature, light and other physical factors which directly influence the organisms in a given area.
2. *Nutrients (material cycling)*¾includes inorganic substances (C, CO_2, H_2O, H_2, P, etc.) and organic substances
3. *Energy circuits*¾includes grazing circuits and organic detritus circuits.

The amount of abiotic materials present in any ecosystem is called *standing state*.

TYPES OF ECOSYSTEMS

Natural Ecosystems

These operate by themselves under natural conditions without any major interference by man. Based upon the particular kind of habitat, these are further divided as:

1. Terrestrial, e.g. forest, grassland, desert
2. Aquatic which is further distinguished as:
   a. freshwater which may be lotic (spring, stream or river) or lentic (lake, pond, pools, ditch, swamp, etc.)
   b. Marine, e.g. sea or ocean (deep bodies) and estuary (shallow bodies).

Artificial Ecosystems

They are also called man-made or man-engineered ecosystems. They are maintained artificially by man where, by addition of energy and planned manipulation, natural balance is disturbed regularly, e.g. croplands such as sugarcane, maize, wheat, rice-fields; orchards, gardens, villages, cities, dams, aquarium and manned spaceship.

Review Questions

1. Brief out the general structure and types of ecosystems.
2. What are natural ecosystems? Give examples.
3. What are artificial ecosystems? Give examples.

5

Freshwater Ecosystem

The study of freshwater habitats is called *limnology*. Fresh water environments include the flowing waters (*lotic environment*) and standing bodies of water (*lentic environment*). The dominant feature of all lotic environments is the continuous movement of water and currents, which cuts the channel, moulds the character of the stream and influences the chemical and organic composition of the water. Water running off the land follows courses of least resistance and develops these as distinct channels by erosion. Young or rejuvenated streams, with a high velocity, erode more than they deposit. Water in slow-moving rivers reflects the characteristic of the terrain; nutrient level and sediment load vary according to region. The slow-moving stream often develops floodplains, meanders, and associated features and terminates in a lake or estuary.

The lentic and lotic ecosystems are fundamentally different from one another because of differences in energy input and flow, and mineral input and circulation.

MICROORGANISMS OF THE AQUATIC (FRESHWATER) HABITAT

1. **Bacteria** Most aquatic bacteria are gram negative because of the following reasons.

- Gram negative envelope is suited for the nutrient diluted environment (oligotrophic).
- Important hydrolytic enzymes are retained in the periplasmic space rather than being excreted and lost to the aquatic environment as in gram positive bacteria.
- Gram negative bacterial cell wall contains lipopolysaccharide (LPS)

that protects the cell against toxic molecules like fatty acids and antibiotics. *Specially* found are *Pseudomonas* sp., *Flavobacterium*, *Achromobacter* and *Alcaligenes*, *Vibrio* sp., *Acinetobacter*, *Staphylococcus* sp. Bacteria in surface region are pigmented.

2. **Molds** Deuteromycetes members, Phycomycetes and Myxomycetes members are found in the marine environments.

3. **Protozoa** Species of Foraminifera and Rodidaxia and many ciliate species are common.

STRUCTURE OF LENTIC HABITAT

Lentic ecosystems include all standing water (freshwater) habitats such as lakes, ponds, marshes, swamps, bogs, etc. Lakes are inland depressions containing standing water. They may vary in size from small ponds of less than a hectare to large lakes covering thousands of square kilometres. They may range in depth from a few centimetres to over 1666 metres.

Ponds however are considered as small bodies of standing water so shallow that rooted plants can grow over most of the bottom. Most ponds and lakes have outlet streams and both are more or less temporary features on the landscape.

The aquatic habitats of lake and pond remain vertically stratified in relation to light intensity, wavelength absorption, hydrostatic pressure, temperature etc. In a lake (Fig. 5.1), there are well recognised horizontal strata which include:

i. Shallow water near the shore forms the *littoral zone*. It contains upper warm and oxygen-rich circulating water layer which is called *epilimnion*. The littoral zone includes rooted vegetation.

ii. *Sublittoral zone* extends from rooted vegetation to the non-circulating cold water with poor oxygen zone, i.e. *hypolimnion*.

iii. *Limnetic zone* is the open water zone away from the shore. It is the zone upto the depth of effective light penetration where rate of photosynthesis is equal to the rate of respiration.

iv. *Profundal zone* is the deep water area beneath limnetic zone and beyond the depth of effective light penetration.

v. *Abyssal zone* is found only in deep lakes, since it begins at about 2,000 meters from the surface.

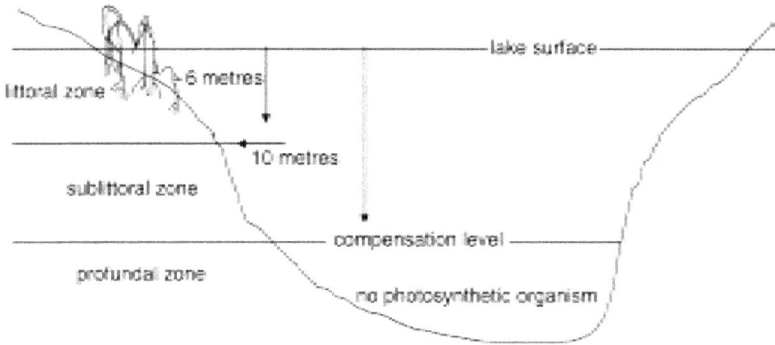

Figure 5.1   *Different zones of a deep freshwater lake*

Ponds have little vertical stratification (Fig.5.2). In ponds, the littoral zone is larger than the limnetic zone and profundal zone. In a small pond the limnetic and profundal zone are not found.

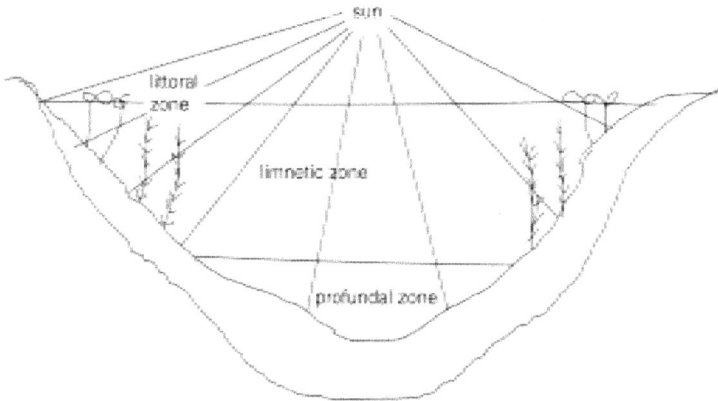

Figure 5.2   *Different zones of a freshwater pond*

BIOTA OF LENTIC HABITAT

Different organisms of the lentic environment can be ecologically classified based on whether they are dependent on the substratum or free from it. Organisms depending on the substratum are called *pedonic* forms and those that are free from it are the *limnetic* forms.

Further, the aquatic organisms may also be classified into the following groups depending upon their sizes and habits.

Neuston   These are unattached organisms that live at the air–water interface. They may consist of microbial forms, plants and animals. Animals and microbial forms that spend their lives on top of the air–water interface, such as water striders are termed *epineuston* while others including insects such as beetles which spend most of their time on the underside of the air–water interface and obtain much of their food from within water, are termed *hyponeuston*.

Plankton   These are forms which are found in all aquatic ecosystems except fast moving rivers. They are small plants and microbial forms whose powers of self-locomotion are so limited that they cannot overcome currents. Thus their distribution is controlled largely by the currents in their ecosystems. The *phytoplanktons* and *zooplanktons* can move a bit, either to control their vertical distribution or to seize prey. Certain zooplanktons are exceedingly active and move relatively great distances considering their small size, but they are so small that their range is still controlled largely by currents and such planktons are also called *nektoplankton*.

Nekton   Nektonic animals are swimmers and are found in all aquatic systems except for fast moving rivers. In order to overcome currents, these animals are relatively large and powerful.

Benthos   The benthos include the organisms living at the bottom of the water mass. They occur virtually in all aquatic ecosystems. The benthos organisms living above the sediment–water interface are termed *benthic epifauna* and those living in the sediment itself are termed *infauna*.

The water surface of a lake or pond contains certain free floating hydrophytes such as *Azolla, Lemna, Wolffia,* etc. The microbial community lives associated with the surface film. The habitat is unusual in many ways since it is subjected to rapid temperature fluctuations, increasing to high light intensity and is very well aerated. Organic matter especially particulate and inorganic nutrients such as phosphates, accumulate on the surface of the film. The habitat is favourable for photoautotrophs. There are increased numbers of bacteria especially *Pseudomonas* and *Caulobacter*.

Neuston is a stressed habitat and microbial growth rate is low. Bacteria in neuston have hydrophobic surfaces and produce extracellular polysaccharides both of which are concerned with adsorption on the surface film.

Bubbles arising through the neuston layer and bursting play a major role in water-to-air transfer of bacteria and viruses. Characteristic autochthonous neuston microbiota include algae, fungi and protozoa. Cyanobacterial species include *Aphanizomenon, Anabaena, Microcystis.* Filamentous fungus include *Cladosporium.* Yeasts are also found. Among protozoan species include *Diffugia, Arcella, Acinata,* etc.

Aquatic life is most prolific in the littoral zone. The littoral zone of a lake remains rich in pedonic flora especially up to the depth to which effective light penetration is possible facilitating the growth of rooted vegetation. Microorganisms exhibit different absorption spectra determining which wavelength of light can be utilised for photosynthesis. Green and purple sulphur bacteria grow at the sediment-water interface below the layers of short wavelength absorbing algal and cyanobacterial growth by utilising wavelength of light not absorbed by the overlying phytoplankton because the purple and green sulphur bacteria obtain electrons from hydrogen sulphide at lower energy loss than water splitting photoautotrophs and thus require lower light intensities for carrying out photosynthesis. Conditions in the euphotic zone (area of light penetration) are favourable for the growth of photoautotrophs.

The bottom of the lake (benthos) represents the interface between hydrosphere and lithosphere. Sedimented organisms in the profundal zone are largely secondary producers and are dependent on the transport of organic compounds from the overlying zone. Particulate nutrients sedimented by gravitational forces concentrate on the surface of the sediment. Growth occurs on the sand grains (sediment) if they are relatively undisturbed by currents and wave action. Flora consists of bacteria and small diatoms (*Fragilaria, Opephora*). There are also some motile organisms like diatoms (*Nitzhia*), cyanobacteria and bacteria. In a fresh water lake, mud found at the sediment, the acidity, water type and the nutrient status controls the microorganisms found there. All muds have diatoms as the major group of microflora. Bacteria are more on the surface and anaerobes are found for some depth in the mud. Oxygen can diffuse only very slowly in the water-filled pore spaces of sediments. Concentration of inorganic nutrients (nitrogen and phosphorus) are important in determining the ability of the habitat to support microbial growth and metabolism. Salt concentration of water and pH also influence the characteristic autochthonous microbes of some lakes with high salt concentration (28%) to develop halophilic (organisms which can grow at high salt concentrations) population (archaea and some algae, e.g. *Dunaliella*).

STRUCTURE OF THE LOTIC HABITAT

Running or moving water or lotic ecosystems include rivers, streams and related environment. They are remarkably variable, ranging in size from large rivers to the trickling small springs.

The lotic habitat is primarily determined by the velocity of the current which can create either slow-moving or fast-moving streams; each has very distinct characteristics. The base of the food chain is dependent on detritus from upstream or from the edges. In slow-moving streams, plant and animal communities largely resemble those found in lentic (lake and pond) habitats. The significant phytoplankton populations that usually exist contribute to a higher rate of primary productivity than that found in fast-moving streams. The level of productivity is dependent upon water temperature and the amount of nutrient input received from the surrounding environment, and therefore subject to seasonal variation. The diversity of consumer organisms varies according to the physical conditions and vegetation. Planktonic populations are relatively high, although not as dense as those found in lakes. In fast-moving streams, there is very little primary production in the open-water habitat, due to the velocity and turbulence of the current. Populations of consumer organisms (mainly particulate feeders) are low. Riffle areas provide valuable habitat for juvenile trout and salmon. Pools are important resting areas for several fish species, including Atlantic salmon. The quality of these areas can be adversely affected when shade trees are removed from the banks.

Successional Sequence

The normally understood process of ecological succession does not apply to open water. In slow-moving streams, the development of habitat depends upon the depositional and erosional characteristics of the river. The fast-flowing, young streams will always be present as the river erodes the landscape. Over time, the young stream will mature into a slow-moving stream, but it can be rejuvenated when a geological obstacle (e.g. a waterfall) is encountered. In mature streams, there is a progressive downstream movement of meanders, leaving shallow or deep pools, backwaters, braided channels and oxbow ponds. There is an associated change in the character of the open water.

BIOTA OF THE LOTIC HABITAT

Vegetation in the lotic open-water habitat consists mainly of phytoplankton found in slow-moving streams. There are no plankton species unique to rivers; those found there originate mostly from backwaters or lakes. Several species of desmids and diatoms are present in slow-moving rivers, although not as abundant as in lakes.

Some zooplankton species and rotifers can be found in slow-moving streams. Their abundance depends on the amount of the predation from invertebrates and small fish. Fish species such as redbelly dace and white sucker, and introduced species such as brown trout are commonly found in slow-moving streams. Fast-moving streams provide excellent habitat for many kinds of fish, including brook trout, Atlantic salmon parr, common shiner, white suckers and yellow perch.

Review Questions

1. Define the following:
   a. Limnology
   b. Neuston
   c. Plankton
   d. Nekton
   e. Benthos
2. Describe the structure of a lentic habitat.
3. Describe the biota of a lentic habitat with special emphasis on neuston.
4. Brief out the characteristic features of a lotic habitat.

6

Marine Ecosystem

About 71% of the surface of the planet is covered by salt water. Physically, the sea has vast inhabitable space, starting from pelagic system to hadal system and littoral to deep (sea) benthic system. The seawater and the ocean floors together form the inorganic marine environment. The diverse groups of marine living forms, their history and propagation are mainly controlled by the chemical composition and various physical properties of the seawater. Also the distribution, concentration, seasonal changes and the movement of seawater and the nature of ocean floors have a major role in the distribution of marine living forms. The water surrounds all the marine organisms and the bodies and internal system of coelenterates, echinoderms, tunicates, etc. Due to the relatively stable nature of physical properties and due to the presence of dissolved salts in seawater, the marine organisms lack specialised regulatory systems or integuments. In this way the organisms themselves are a part of the dynamic environment.

PHYSICAL AND CHEMICAL CHARACTERISTICS OF THE MARINE ENVIRONMENT

In general, water is most essential for the maintenance and activities in all life forms. Water is an important component of cell constituting about 80% of the weight of protoplasm. For the photosynthetic process in autotrophic plants, water itself is an important raw material. It is a universal solvent, which carries the necessary gases like oxygen and carbon dioxide and also the growth regulating minerals in dissolved condition. In aquatic environments, there is no problem of dessication and consequently no specialisation in the organisms. In view of high transparency of water, photosynthesis is possible even at a relatively deeper level. The sea water can change from acid to alkaline condition and vice versa. This buffering nature is useful to organisms because an abundant supply of carbon dioxide is necessary for plants for photosynthesis. In case of alkaline condition, the construction of shells by marine organisms by using calcium carbonate is enhanced. The lower specific gravity of sea water is most beneficial to marine organisms. As the sea water contains large number of salts it is a most suitable environment for living cells. Further it has been found that the ratio of total salt content of seawater is almost same as that of body fluids of many invertebrates.

In general, the marine environment offers a wide range of living conditions. The salinity ranges from very dilute estuarine condition to as high as 37 %. Varying light conditions exist with brilliant sunlight at the surface waters to no light in the deep waters. Likewise, the pressure varies from 1 atmosphere at the surface to 1000 atmospheres at greater depths. These gradients of environmental parameters are favourable to a number of sensitive animals. The more fluctuations of environmental features are especially encountered in coastal areas due to their peculiar physiographic characters. The water movement/circulation is useful in the oxygenation of subsurface water, for the dispersal of metabolic wastes and plant and animal (growth) nutrients, and also for the disposal of spores, eggs, larvae and even adults.

STRUCTURE OF THE SEA

The littoral (eulittoral) zone occurs at the seashore (Fig. 6.1). This zone is subjected to alternate periods of flooding and drying at high and low tides respectively. The *sublittoral* zone extends from the low tide mark to the edge of the continental shelf. This region is also called as the *neritic*, or near shore zone. The term *pelagic* is used to designate open water or the high sea and includes portions of the neritic and the entirety of the oceanic province. The benthos or benthic region is the bottom, regardless of the overlying zone.

The *benthic region* begins at the *intertidal zone* (*littoral zone*) and extends downwards. The continental shelf is a gently sloping benthic region that extends away from the land mass. At the continental shelf edge, the slope greatly increases. The continental slope, also known as the *bathy region*, drops down to the sea floor. The *deep-sea floor* is known as the *abyssal plain* and usually lies at about 4000m. The ocean floor is not flat but has deep ocean trenches and submarine ridges. The *deep ocean trenches* are called as the *hadal region*.

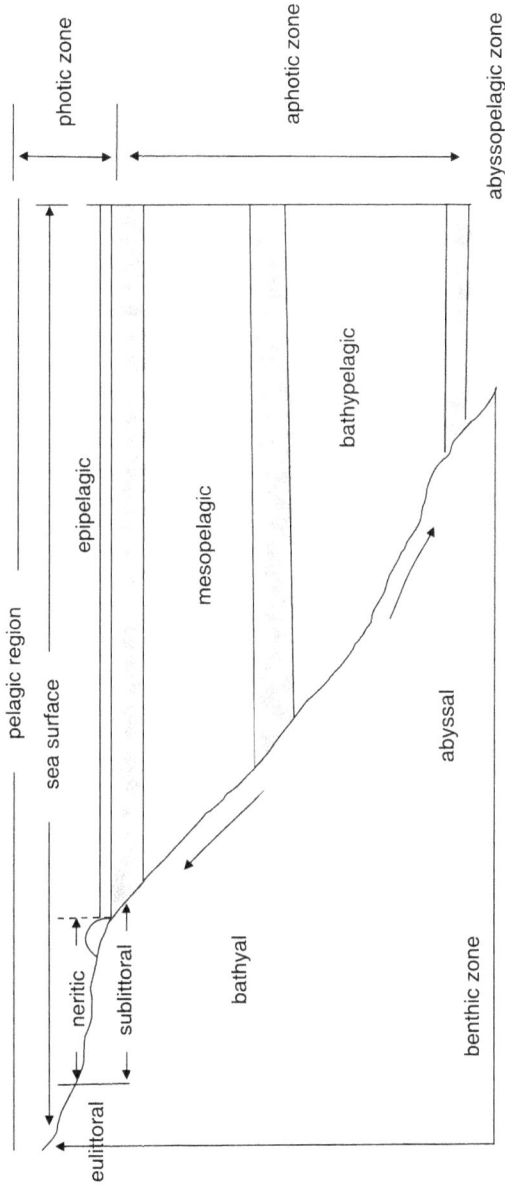

Figure 6.1   *Zonation of the marine environment (sea)*

UPWELLING

Upwelling describes a condition in which the cooler, nutrient-rich waters of the ocean beneath the thermocline are drawn upwards to replace the warmer, surface waters which have been displaced by surface winds. It is also the vertical upward movement of a fluid due to density differences or where two fluid masses converge, displacing fluid upward. Upwellings are caused by strong seasonal winds moving surface coastal water out from the coast and leaving a space that the upwelling fills in. Many marine plants and animals live off this nutrient-rich water. The thin horizontal layer of water riding on top of the ocean that is affected by wind is the Ekman layer. Ekman transport causes surface waters to diverge or move away from the coast and deeper (often cold and nutrient-rich) water to be brought to the surface. Because winds blowing on the sea surface produce an Ekman layer that transports water at right angles to the wind direction, any spatial variability of the wind, or winds blowing along some coasts, can lead to upwelling as seen in Fig. 6.2.

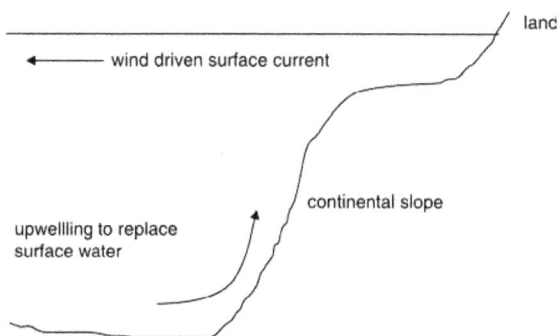

Figure 6.2 *Formation of upwelling current*

Fig. 6.3 shows a sketch of Ekman transport along a coast leading to upwelling of cold water along the coast. The water transported offshore must be replaced by water upwelling from below the mixed layer (Fig. 6.3a). North winds along a west coast in the northern hemisphere cause Ekman transports away from the shore (Fig. 6.3b).

The winds produce a mass transport away from the shore everywhere along the shore. The water pushed offshore can be replaced only by water from below the Ekman layer and, this is upwelling (Fig. 6.3b). Because the upwelled water is cold, the upwelling leads to a region of cold water at the surface along the coast.

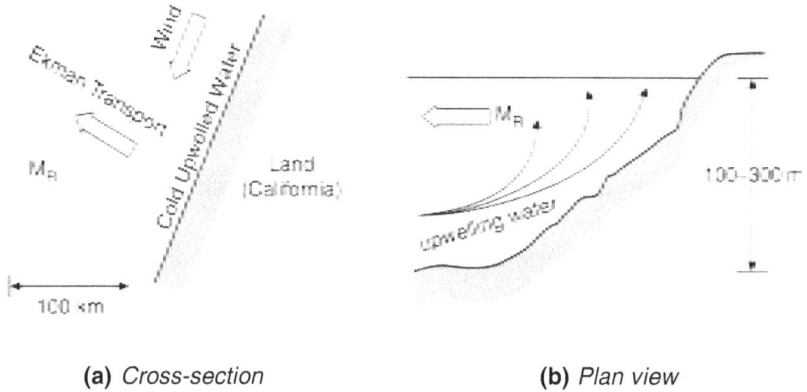

(a) *Cross-section*   **(b)** *Plan view*

Figure 6.3   *Sketch of Ekman transport*

Upwelled water is colder than water normally found on the surface, and it is richer in nutrients. The nutrients fertilise phytoplankton in the mixed layer, which are eaten by zooplankton, which are in turn eaten by small fish. The small fish are eaten by larger fish and so on. As a result, upwelling regions are productive waters supporting the world's major fisheries.

Atmospheric winds generate horizontal currents that move around the ocean's surface. Wind can also generate vertical water motions in the processes of upwelling and downwelling. When wind blows over water, the surface water does not move directly in front of the wind but moves about 45 degrees towards the right of the wind's motion in the northern hemisphere (Fig. 6.3a). This process is called Ekman transport and is a result of the Coriolis effect. In the southern hemisphere, surface water is deflected to the left of the wind's motion. Where winds cause the surface water to move away from a coastline or to diverge from another surface water mass, deeper water will move up to the ocean surface, creating an *upwelling current*. Where winds cause the surface water to move towards a coastline or to converge with another water mass, the surface water will try to move downward to create a *downwelling current*. For example, northerly winds are common in summer along the California coast. Winds moving from north to south cause surface water to move towards the west, away from the coastline. Upwelling currents are created, which bring deeper, colder water to the surface.

While surface waters are usually depleted of nutrients such as phosphates and nitrates that are crucial for plant growth, deeper waters have high concentrations of these nutrients. Upwelling replenishes the surface layers with the nutritional components necessary for biological productivity. Regions of upwelling are among the richest biological areas of the world.

The importance of upwelling can be well understood from the following facts.

- Upwelling enhances biological productivity, which feeds fisheries.

- Cold upwelled water alters local weather. The onshore weather in regions of upwelling tend to have fog, low stratus clouds, a stable stratified atmosphere, little convection, and little rain.

- Spatial variability of transports in the open ocean leads to upwelling and downwelling which in turn leads to redistribution of mass in the ocean, which leads to wind-driven geostrophic currents via Ekman pumping.

Significance of Upwelling

Typically, upwelling systems are dominated by diatoms, which are capable of fixing substantial amounts of carbon. Organic matter is ultimately removed from the euphotic zone by a range of processes including turbulent mixing, detrainment, and downwelling. However, biologically mediated processes often dominate vertical transport through mass aggregation and sedimentation during diatom blooms; nutrient depletion may play an important role in enhancing aggregation during periods of upwelling relaxation. Ingestion of phytoplankton by vertically migrating zooplankton is another mechanism for transporting carbon to depth. The production rate of zooplankton is an important factor in the health of coastal fisheries. The carbon in phytodetritus and faecal pellets sinks to the seabed on continental shelves and slopes, where it fuels benthic productivity and leads to temporally variable bioturbation and diagenetic chemical reactions. Nutrients released into pore fluids can return to the water column through seabed erosion and biological irrigation, thus helping to charge deep waters for the next upwelling event.

DOWNWELLING

The vertical downward movement of a fluid due to density differences or the downward displacement of fluid when two fluid masses converge is called downwelling. In the ocean, it often refers to where Ekman transport causes surface waters to converge or impinge on the coast, displacing surface water downward and thickening the surface layer.

BIOTA OF THE SEA

The marine environment is subjected to many changes including change in pressure, light intensity, turbidity, dissolved salts, etc. Based on these factors bacteria in the marine environment can be categorised into three groups:

Barotolerant   They grow best at normal atmospheric pressure but can grow between 0–400 atm.

Moderate barophiles   These grow at 400 atm and also grow at 1 atm.

Extreme barophiles   These are bacteria growing only at higher pressure 6000–11000 m depth at a pressure of 600–1100 atm where the temperature is 2–3°C. These bacteria live in the guts of deep sea invertebrates like amphipods and holothurians.

As with freshwater environment, the uppermost layer of the marine ecosphere is the surface tension layer, where the marine ecosphere interfaces with the atmosphere. The seawater-air interface is the habitat for the pleuston, the marine equivalent of the neuston, which includes bacterial and algal inhabitants. *Pseudomonas* and various pigmented genera such as *Erythrobacter*, *Erythromicrobium*, *Protominobacter* and *Roseobacter* are major bacterial population. A higher proportion of pleuston bacteria have been reported to utilise carbohydrates than do bacteria in the underlying water. Population of primary producers, including cyanobacteria, diatoms and Phaeophycophyta are sometimes found in the pleuston layer.

General Characteristics of Sea Biota

The populations of the sea may be grouped under three major categories, viz. plankton, nekton and benthos.

While the plankton and nekton are found in the pelagic environments, the benthos inhabit the benthic region. Benthos are sessile, creeping or burrowing organisms which spend most or whole part of their life on or inside the benthic substrata. Organisms of this group are found distributed from the highest high tide level to the deep bottom of the sea. The benthic forms are represented by (a) sessile animals like sponges, barnacles, mussels, oysters, corals, worms, seaweeds, benthic diatoms and (b) creepers like crabs, lobsters, some copepods, amphipods, crustaceans, protozoans, snails, fishes, etc. and (c) burrowers like clams, worms, echinoderms, etc.

The group nekton (swimming) comprises active swimming animals is found in the pelagic part. Most of the adult squids, fin and shellfishes, and mammals like whales come under this category. These are larger animals provided with some swimming organs with which they can swim and move against the water currents, waves, etc. The group plankton (wandering) comprises minute, microscopic plant and animal components of the pelagic system, which are passive drifters. These floating creatures are carried away by the water currents. The two main divisions of plankton are phytoplankton and zooplankton. While the former include minute floating plants, the latter is represented by diverse permanent animal plankton and also the eggs and larvae of swimming and benthic organisms.

Nutrient availability is extremely important in determining microbial productivity in marine habitats. Recycling of mineral nutrients is very slow in the pelagic environment. Dead organisms from the euphotic epipelagic zone (zone of effective light penetration) sink into the great depths of the bathypelagic and ultimately the benthic zone. They carry with them essential nutrients, mainly nitrogen and phosphorus that are liberated in the perpetual darkness of the deep ocean. From here they are returned to the surface water by upwelling currents at extremely slow rates. Upwelling phenomena often occur along the continental slope, caused by surface currents running rapidly away from the shore and being replaced by deeper, nutrient rich water as shown in Fig. 6.2.

Marine plankton is the major source of organic matter and comprises numerous species of diatoms, cyanobacteria, dinoflagellates, chrysomonads and chlamydomonads. Phytoplanktons (phyto-plants, plankton-wandering) are otherwise called free-floating plants. A common planktonic alga is *Synecoccus*. *Picocyanobacteria* may represent 20–80% of the total planktonic biomass. These act as a source of food for marine fishes.

Planktonic algae, under certain environmental conditions, may grow into an enormous population with resultant discolouration of water, a condition referred to as bloom. For example, the characteristic colour of 'red sea' is associated with blooms of pigments.

Bacterial population depends on the distribution of phytoplankton since the latter contributes organic substance and solid surfaces for bacterial

Table 6.1 *Marine microbes*

The size of marine microbes ranges from 0.3μm to about 150μm.

Archaeobacteria	Autotrophic eubacteria	Chemoheterotrophic eubacteria	Eukaryotes
Chemoautotrophs	Photoautotrophs	Gram-positive	Photoautotrophs
Methanogens	Anoxygenic photosynthesis	Endospore-forming rods and cocci	Microalgae Chemoheterotrophs
Thermoacidophiles	Purple and green photosynthetic bacteria (Order Rhodospirillales)	Non-spore forming rods Non-spore forming cocci (Family Micrococcaceae)	Protozoan flagellates
Chemoheterotrophs	Oxygenic photosynthesis Cyanobacteria (Order Cyanobacteriales) Prochlorophytes (Order Prochlorales) Chemoautotrophs Nitrifying bacteria (Family Nitrobacteriaceae) Colourless sulphur oxidising bacteria Methane oxidisingbacteria (Family Methylococcaceae)	Actinomycetes (Order Actinomycetales) and related organisms. Aerobic Gram-negative rods and cocci (Family Pseudomonadaceae) Facultative aerobe (Family Vibrionaceae) Anaerobic sulphur reducing bacteria Gliding bacteria (Order Cytophagales and Beggiatoales) Spirochaetes (Order Spirochaetales) Spiral and curved bacteria (Family Spirillaceae) Budding and/or appendaged bacteria Mycoplasmas (class Mollicutes)	Amoebae Ciliates Fungi Higher Fungi Ascomycetes Deuteromycetes Basidiomycetes Lower fungi (Class Phycomycetes)

aggregation. Growth of psychrophilic and halophilic microbes are common. Among the psychrophilic organisms are the luminous bacteria that can produce light in the presence of oxygen. These bacteria exist in symbiotic association with marine animals.

BACTERIAL FLORA OF THE SEA

They are more abundant near the shore particularly in polluted areas. They are sparsest at great depths in open oceans. The generation time of bacteria ranges from less than an hour to months or even longer. The shortest generation time reported is 9.8 minutes for *Pseudomonas natrieganus*. The most characteristic feature of marine bacteria is their capacity to survive and grow in the sea. In coastal regions (where river water enters the sea) there is no sharp distinction between freshwater forms and marine forms. Excepting spores, most freshwater forms perish within a few hours under oceanic conditions. Certain bacterial endospores or fungal spores seem to survive in a dormant state for prolonged periods.

The, typical characteristic of marine bacteria are their small size ($0.1 \ \mu m^3$) and low DNA content. They are *obligately oligotrophic*. Most oceanic bacteria grow more slowly and form smaller colonies on submerged slide (a technique used to study the occurrence of marine flora). Oceanic bacteria are more proteolytic than soil bacteria that are saccharolytic and more often they are *facultatively aerobic*. Most of the marine bacteria liquefy agar.

Most bioluminescent bacteria come from the sea including the saprophytic, symbiotic and commensal species. Oceanic flora can survive extremely low concentrations of nutrients ($20 \ \mu g/l$) hence they are most often associated with algal surfaces or detrital particles which offer nutritional advantage.

Most of the oceanic bacteria grow at $0°C$ (optima is12–25°C). Heating sea samples for 10 min at 30–40°C kills 80% of the bacteria. Most of the sea flora are barotolerant and grow at pH of 7.2–7.6 at the surface and pH 6.4–9.4 in marine sediments. Marine bacteria have a highly specific need for sodium and chlorine. Some marine bacteria have multiple membranes surrounding their cells. Exposure to fresh water, disrupts these membrane layers causing a loss of viability. A high percentage of marine bacteria have a tendency to grow attached to solid surfaces (hence submerged slide technique is used extensively to study the occurrence of marine flora). Numerous species are attached to plankton or to larger organisms.

The highest biomass of microflora in marine waters is normally seen near the surface and they decrease with depth. Some microflora growing in the surface go more downwards often attached to sediment particles. They provide food sources to organisms growing in pelagic habitats. High numbers of heterotrophic bacteria and cyanobacteria are transported into the deep sea attached to rapidly sedimenting particles. Hence, there is every possibility of genetic exchange between populations previously assumed to be genetically isolated.

Most marine bacteria are gram negative and motile. For example, *Pseudomonas, Vibrio* are dominant genera followed by *Flavobacterium, Alcaligenes* and *Microcystis. Actinomycetes* are also found in marine waters. Some gram positive bacteria like *Bacillus* are found in marine sediments. Below the sediments, anaerobic bacteria are composed of the autochthonous microorganisms including *Desulfovibrio* and methanogenic bacteria. Chemolithotrophs include *Nitrosomonas, Nitrosococcus, Nitrospira* and *Nitrobacter.*

Fungal flora of the sea   Yeasts are more common, e.g. *Torulopsis, Candida, Cryptococcus, Saccharomyces, Rhodosporidium*.

Algal flora of the sea   The algal flora usually belong to Chlorophyta, Phaeophyta, Rhodophyta, Cyanophyta. e.g. *Sargassum, Fucus*. Chlorophyta and Chrysophyta members are prominent among phytoplankton found in the upper regions of surface (0–50 m). Planktonic diatoms are common forms, e.g. *Trichodesmium*. Occasional blooms of Phaeophycean members are common in the oceans causing the 'red tides' (ocean becomes red brown in colour). The toxins produced by some of these dinoflagellates kill fishes and other marine animals.

Some of the dinoflagellates produce neurotoxins which accumulate in shell fishes that feed on them. When ingested by humans such tainted sea food can cause paralytic shell fish poisoning (PSP).

Protozoan flora of the sea   They are important components of marine zooplankton. They can tolerate up to 10% NaCl concentration. They include flagellates, rhizopods and ciliates, e.g. *Radiolaria, Acantharia*. They occur in the deeper layers of the sea. Marine protozoa graze on bacteria, phytoplankton and smaller forms of zooplanktons. Grazing provides a critical link in the marine food web between very small primary producers and the higher members of the marine food web.

Functions of Marine Flora

1. They cause spoilage of fish and other marine food products. *Clostridium botulinum* is quite commonly present in fishes, clams and oysters.

2. Harmful bacteria bring about the deterioration of various manmade structures such a fish nets, ropes, soil cloth, etc.

3. The most important activity of bacteria in the sea is the mineralisation, modification and synthesis of organic matter (carbohydrates, proteins, hydrocarbons, chitin).

4. Both chemosynthetic and photosynthetic autotrophs occur in the sea, thus increasing the primary productivity of the sea.

5. More than 100 species of microbes have been shown to fix nitrogen found in sea water/bottom deposits.

6. Certain chemical transformations caused by bacteria tend to affect the pH of the environment.

pH may be increased by (a) formation of ammonia, (b) reduction of nitrates, nitrites and sulphates, (c) oxidation or decarboxylation of organic acids, or (d) reduction of carbon dioxide.

pH can be decreased by (a) bacterial oxidation of ammonia to nitrite and nitrate, (b) oxidation of sulphur and hydrogen sulphide to sulphate, (c) formation of organic acids/carbon dioxide from carbohydrates, or (d) liberation of phosphates from phospholipids.

Generally these activities go hand in hand thus maintaining neutrality. Sometimes either one may predominate as in the sediments.

Marine bacteria are believed to promote the transformation of petroleum from marine sediments. In highly reducing environments (anaerobic), bacteria tend to convert organic matter partly into carbon compounds relatively richer in hydrogen and poorer in oxygen, nitrogen, sulphur and phosphorus. This results in the formation of carbon compounds that are more like petroleum.

Some marine bacteria can oxidise oil by degrading the oil; they can thus reduce danger of oil pollution in the marine environment.

Review Questions

1. Define the following:
   a. Upwelling
   b. Downwelling
   c. Pleuston
2. Brief out the structural aspects of marine environment.
3. Discuss the phenomenon and significance of upwelling.
4. Describe in detail the biota of the marine environment giving their general characteristics.
5. Discuss bacterial flora of the sea.
6. Brief out the functions of marine microflora.

7

Estuarine Ecosystem

An estuary is a semi-enclosed body of water with variable salinity intermediate between salt and fresh water. Examples of estuaries are river mouths, coastal bays, tidal marshes and water bodies behind barrier beaches. Estuaries are among the most naturally fertile waters in the world (Odum, 1989). Their high productivity results from their unique juxtaposition at the edge of the continent. Nutrients from four sources contribute to the productivity of estuaries:

1. fresh water flowing off the land;
2. tidal exchange with the ocean;
3. the atmosphere; and
4. the recycling of material from the estuarine bottom sediments.

The most important nutrient is nitrogen¾a component of all proteins. Phosphorus, silica, and other compounds in lesser amounts also serve as nutrients to living things in the estuary.

The estuary functions as an efficient nutrient trap that is partly physical and partly biological. Three major forms of photosynthesising organisms play key roles in maintaining high productivity by exploiting nutrient sources.

i. phytoplankton suspended within the sunlit zone of the water column;

ii. benthic microflora which are microscopic plants living on the sediment surface wherever sufficient light reaches the bottom; and

iii. macroflora or rooted plants and rootless algae growing in shallow water and along the shoreline.

These plants are the foundation of complex food webs and provide structural habitats which create natural habitat for most coastal shellfish and finfish. Physical processes contribute to the acquisition and transformation of nutrients by living things. For example, the importation of nutrients and exportation of waste products to and from the estuary are subsidised by gravitational energy in the form of streamflow and tidal exchange. As a result, the estuary becomes a productive seafood factory.

One component of the estuary is shallow salt marshes that play an important role for organisms. Salt marshes are regularly rinsed by the ebb and flow of the tides. Waste products are diluted and removed and nutrients are brought in from the land and the sea. The materials of streams flocculate as the mixture of clay-sized particles and algae meet the sea waters. They settle down, leaving the water clear. In the salt marsh, plants need not expend energy in collecting minerals. They are assisted in obtaining minerals from many organisms. Clams, oysters and mussels trap nutrients as they feed, and deposit the wastes as pseudo-faeces. In these nutrient-rich waters, plants grow, die, and cycle rapidly as they are consumed by the inhabitants of the estuary. The major energy flow is by way of detritus food chain rather than the grazing food chain.

BIOTA OF ESTUARIES

In estuaries, the plants of different groups are present, including phytoplankton, benthic diatoms, bacteria and fungi and larger macrophytes. Though estuaries are rich in their nutrients phytoplanktonic organisms are not abundant. This is due to the reduction in light penetration as a result of turbidity. Most of them are marine in origin and fresh water phytoplankton is poorly represented.

The important estuarine phytoplanktonic organisms are diatoms and dinoflagellates. The diatoms are *Skeletonema costatum*, *Parallia sulcata*, *Chaetoceras debills*, etc. *Ceratium furga* and *Ceratium buceros* are the dinoflagellates of the estuaries. The benthic diatoms are bottom dwelling organisms which are the primary producers and serve as food for a variety of estuarine animals. *Euglena obtuse* is quite common in the bottom of the estuaries.

Both aerobic and anaerobic bacteria inhabit the estuarine water and also the bottom. Sulphate reducing bacteria are found in the mud. Generally all types of bacteria decompose the organic substances of the estuaries. The fungi such as *Mucor* and *Pencillium* also decompose the organic materials of the estuaries.

The macrophytes are larger plants which are also found in the estuaries. Brown algae like *Fucus, Cladophora* and *Vaucheria,* Chorophyceae and Xanthophyceae are the common algae of the estuaries. Besides, eel-grass are the typical vegetation of the estuaries.

The planktonic organisms of the estuaries are not true estuarine forms, but are brought into the estuaries from the sea by the tides. They are again carried out on the ebb. Thus, there is only a temporary plankton whose stay in the estuary is brief, being limited to the duration of a single tide. Planktonic organisms of fresh water origin may be brought into the estuary by the river which are finally carried to the sea and are perished. The fresh water planktons such as *Daphnia, Bosmina, Holopedium, Cyclops* and *Diaptomus* are usually found in the estuarine water.

In estuaries with a long flushing time and a fairly stable salinity gradient there is the possibility of a permanent plankton.

Moreover estuaries possess distinct blooms such as red tides of large blooms of red pigmented dinoflagellates such as species of *Gonyaulax* and *Gymnodinium.*

Review Questions

1.   Discuss estuary as an ecosystem.

2.   Give an account of biota of estuaries.

8

Mangrove Ecosystem

The word 'mangrove' originates from 'mangue', a combination of the Malay, Spanish, French or Portuguese for 'wood' and 'grove' English for 'a small wood'. Mangroves are the special types of habitats found in the coasts of subtropical and tropical countries. They are bordered by shallow sea water and thus protected from direct wave action. Mangroves are defined as a type of coastal woody vegetation that fringes muddy saline shores and estuaries in tropical and subtropical regions.

CHARACTERISTIC FEATURES OF MANGROVES

- A unique characteristic feature of all mangroves is the muddy bottom. The bottom is muddy due to the absence of wave actions and strong water currents. The muddy bottom is exposed during low tides.

- The mangroves are always shallow in nature, and rarely show high depth at certain regions.

- Generally mangroves are developed on coasts where the atmospheric temperature is never below 20°C even during the winter season.

- A high rainfall is necessary for the development of mangroves.

- The water of mangrove shows high salinity due to the concentration of salts such as sodium chloride, and magnesium sulphate. Besides, the water shows dissolved ions, manganese and molybdenum.

- They are water logged or stagnant with high nutrients brought in by the incoming fresh water. The muddy bottom also contains rich organic matter.

ZONATION

On the basis of salinity, five zones of mangrove distribution are considered. These are the *euhaline, polyhaline, mesohaline, oligohaline* and *limnetic zones*. In India, the west coast is characterised by the rocky substratum and hence absence of mangroves in the mouth region. On the other hand, the same euhaline zone along the estuaries of the east coast, which is a delta region, shows the presence of luxuriant mangrove forests as observed in the Gangetic, Mahanadi and Godavari deltas.

In the *euhaline zone*, the salinity ranges from 30 ppt to 40 ppt. Wave action is maximum and the gradient is not steep; thus sediment deposition will take place in the region of confluence which is known as delta. Otherwise, the entire sediment load will be washed into the sea.

The *polyhaline region*, characterised by salinity range of 18 ppt to 30 ppt, has a low wave action and substratum is sandy clay.

The *mesohaline region*, with 5 ppt and 18 ppt salinity, has silty clay bottom and feeble wave action.

In the *oligohaline zone*, the salinity is reduced further to 0.5 ppt to 5ppt as a result of more freshwater influx. Its substratum is silty.

The *limnetic zone* is almost freshwater with occasional intrusion of brackish water at the highest high tides. The salinity in this area is less than 0.5 ppt and the substratum is full of gravel and coarse grains of sand.

Different mangrove species, according to their salt tolerant ability and substratum preference occupy different zones along the estuaries. The deposition of sediment at the mouth of the major estuaries like Ganga, Mahanadi, Godavari, etc. is quite common. During this process, the fine alluvial silt is deposited in the form of shallow islands in the mouth region. Slowly the new mangrove formation develops following a typical succession pattern. After the growth of pioneer species, the dominant or climax vegetation of mangrove species takes over and flourishes.

These new mangrove formations are very sensitive to the flood waters during rainy season. Depending upon the flood, the rate of erosion in the mangrove soil varies. During heavy to very heavy floods, these new mangrove formation may be completely washed out. During the rainy season, the rate of erosion is higher while during the non-monsoon season the accretion is higher. The effect of accretion and erosion seem to be more prominent along the east coast than on the west coast of India.

BIOTA OF MANGROVE

The succession of mangrove is dependent on the available seeds or propagules, their size or length and the tidal fluctuation. Seeds of grass, sedge or *Excoecaria agallocha*, which are minute in size, will always establish themselves at the uppermost limit of the intertidal region. At the same time, seedlings of taxa like *Rhizophora*, *Kandelia*, *Ceriops* or *Bruguiera* will be established according to their floating height.

The major ecological role of mangroves is the stabilisation of the shoreline and prevention of shore erosion. The dense network of prop roots, pneumatophores and stilt roots not only give mechanical support to the plant, but also trap the sediments. The rate of sedimentation or accretion is generally much higher in these estuaries lined with mangroves.

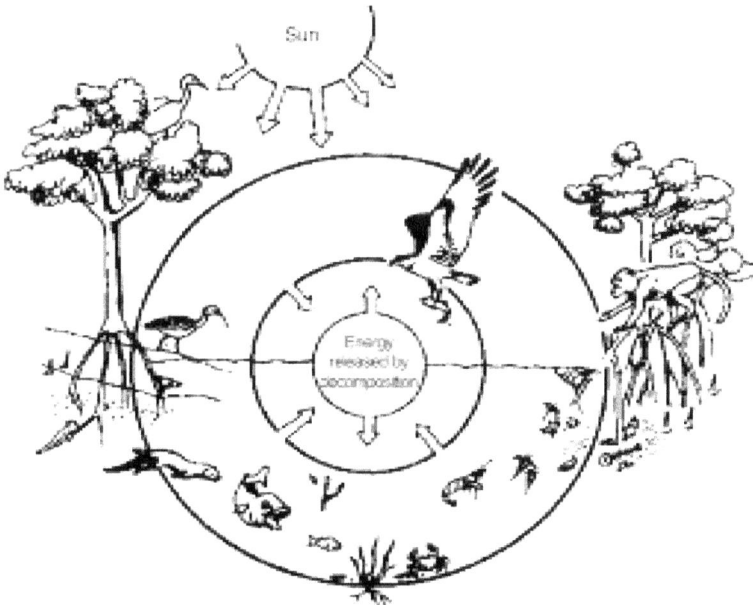

Figure 8.1 *Mangrove ecosystems*

The second important ecological role of the mangroves is the detritus, which helps in feeding and provides breeding and nursery grounds for the juveniles of many commercially important shrimps and fishes. Major primary production in the mangrove ecosystem is from the trees. However, only a fraction of this production is consumed by herbivores. The remainder enters the mangrove water as litter fall. The decomposition of this litter fall produces detritus, which in turn is colonised by heterotrophic microorganisms, thus enhancing its nutritive value. The detritus, besides forming a food source for suspension and deposit feeders, is also consumed by the juveniles of a variety of bivalves, shrimps and fishes, which migrate into the mangrove environments in their life cycle for better feeding and protection. There is a direct correlation between the extent of mangrove forests along a coastline and the fishery as well as shrimp catches from the coastal waters adjoining the mangroves, thus demonstrating the importance of mangroves for sustaining coastal fisheries.

In mangrove forests, the floral elements responsible for the photosynthesis under brackish water condition are of different types, i.e. angiospermic flora, phytoplankton and marine algae. These elements contribute mainly to the primary productivity. Apart from this, faunal elements like zooplankton are responsible for secondary productivity and benthic animals for tertiary productivity.

The primary productivity of phytoplankton or plankton production in mangrove environment represents only that of surrounding waters, i.e. the estuarine water flowing in and out of the mangrove area.

Other Interrelated Ecosystems

There are certain other ecosystems interrelated with mangrove, which include:

Benthic community   Major groups represented by the benthic organisms are molluscs, crustaceans, echinoderms, hydroids, actinarians, planarians, nematodes, polychaetes and larval forms of several other organisms.

Pelagic community   The mangrove water usually rich in detritus, is highly suitable for fishing. The major fishery resources found in these waters are detritivorous species of fishes, crabs, crustaceans and molluscs. Roughly about 60% of India's coastal marine fish species is dependent on the mangrove–estuarine complex. The annual landing of the fish from the Hoogly-Matlah estauries of Gangetic Sundarbans have exceeded 10,000 tons.

Some of the most common fishes in Indian mangrove waters are species of *Liza, Mugil, Lates, Polynemus, Sciaena, Setipinna, Pangasium, Hilsa, Ilisha* and *Atroplus*. Prawns are represented by the species of *Penaeus* and *Metapenaeus*, while the crabs are represented mainly by *Scylla serrata*. The molluscs of mangrove waters are mainly represented by *Crassostrea* sp., *Mytilus* and clams.

In the upstream regions, giant prawns like *Microbrachium rosenbergii* are also found in large quantities.

The important phytoplankton of the mangroves are *Pleurosigma, Bacillaria, Navicula, Thalassiotirix, and Biddulphia* (diatoms). Dinoflagellates such as *Ceratium, Peridiniem* and *Procentrum* are also quite common in the water of mangroves.

Review Questions

1. Discuss mangroves as an ecosystem briefing on its characteristic features.

2. Give an account of biota of mangroves.

3. Discuss the role of mangroves in any habitat.

PART III

MICROBES AND ATMOSPHERE

Relationship between Microbes and
  Atmosphere

Sampling of Air

Potential Hazards in Laboratory

Techniques Airborne Diseases

Air Sanitation

Air Pollution

9

Relationship between Microbes and Atmosphere

The ecosphere or biosphere constitutes the totality of living organisms on earth and the abiotic surroundings they inhabit. It can be divided into atmo-, hydro- and litho-ecospheres to describe the portions of the global expanse inhabited by living things in air, water and soil environments, respectively. Microorganisms live within the habitats of these ecospheres. Each habitat has a set of physical, chemical, and biological parameters that determine the microbial populations that may thrive there. As a result of natural selection forces, characteristic communities develop within each habitat. In some cases, particularly in extreme habitats such as salt lakes and thermal springs, the indigenous microbial populations exhibit adaptations to their physical and chemical surroundings that permit their survival. In other habitats, intense competition dictates which populations survive and become the autochthonous members of the communities living there.

The vast majority of airborne pathogens are uniquely adapted for spreading in indoor environments. The conditions of temperature, humidity and protection from sunlight and from oxidants which man controls for his own comfort serve also to protect pathogens during their exposed and vulnerable period when they transmit themselves from one person to the next. Most airborne pathogens die out rapidly in outdoor air but as individual species they depend entirely on man and his indoor environments for their propagation.

The term 'air pollution' is therefore applied when there is an excessive concentration of foreign matter in the outdoor atmosphere which is harmful to man or his environment. Air pollution is a growing menace to health throughout the world. The problem of air pollution was first brought to a sharp focus when air pollution epidemics took place in the developed countries during 1940s when some 4000 people died within 12 hours. These epidemics aroused public interest and stimulated the health authorities to take steps to ensure clean air.

The different layers of the atmosphere are troposphere, stratosphere and ionosphere. Air is the simplest ecosystem since it consists of a single gaseous phase apart from condensed water vapour and dust. Atmosphere consists of 79% nitrogen, 21% oxygen and 0.034% carbon dioxide. It is saturated with water vapour at varying degrees. For the most part, the chemical and physical parameters of the atmosphere do not favour microbial growth and survival. Temperature decreases with increasing height in the troposphere (region nearest the earth's surface interfaces with both hydrosphere and lithosphere). At the top of the troposphere, temperature is –43°C to –83°C which is below the minimal growth temperature for microbes. With increasing height in the atmosphere, the atmospheric pressure declines and concentration of available oxygen decreases to a point that prevents aerobic respiration. The concentration of organic carbon is low which is not sufficient enough to support heterotrophic growth; available water is scarce limiting even the possibility of autotrophic growth of microbes in the atmosphere. Microbes in the atmosphere are exposed to high intensity of UV radiation which increases with height as the atmosphere thins and offers less shielding from UV radiations and causes lethal mutations and death of microbes.

Stratosphere   It contains a layer of high ozone concentration which absorbs the UV light protecting the earth's surface from excessive UV radiation. The stratosphere represents a barrier to the transport of living microbes to or from the troposphere. Organisms in this zone are thus transported slowly and are exposed for a prolonged period to the prevailing concentration of ozone and high UV light intensities. Only microbes shielded from these conditions in the stratosphere could survive passage out of earth's atmosphere.

ATMOSPHERE AS A HABITAT

Even though atmosphere is a hostile environment for microbes, there are substantial number of microbes in the lower troposphere, where because of thermal gradients there is rapid mixing of air. Movement of the air represents a major pathway for the dispersal of microbes. Several bacterial, viral and fungal diseases are spread through the atmosphere.

Temporary locations in the troposphere may provide habitats for microbes. Clouds possess concentrations of water that permit growth of microbes. Light intensities and CO_2 concentration in cloud layers are sufficient to support growth of photo-autotrophs.

In industrial areas, there may even be enough concentration of organic chemicals in the atmosphere to permit growth of some heterotrophs. But conclusive proof is lacking and the practical importance of such life appears to be negligible.

Although many microbes that grow in the hydrosphere or lithosphere can become airborne, there are no known autochthonous atmospheric microbes. During dispersal, aquatic and soil microbes may enter and pass through the atmosphere before reaching other favourable aquatic or terrestrial ecosystems. Dispersal through the atmosphere ensures continued survival of many microbes. Microbes in the atmosphere occur as spores/soredia/ cysts and other vegetative resistant structures. Spores whose primary function is dispersal are known as *xenospores*. Fungi, algae, some protozoa, actinomycetes and lichens produce spores that are present in the atmosphere. Viruses are transported through the atmosphere as inactive particles that are functionally equivalent to the dormant spores of living microbes. .

TRANSPORT OF SPORES

The low metabolic rates of spores mean that they do not require external nutrients and water to generate sufficient energy for maintenance over long periods. Spores are produced in very high numbers. For example, fungi produce in excess of 10^{12} spores/fruiting body/year. A large percent of spores do not survive transport in the atmosphere. Some spores have extremely thick walls which protect them from severe dessication. Some spores are pigmented which adds protection from exposure to damaging UV radiation. Their small size and low density permit them to remain airborne for long periods before they sediment from the atmosphere. Spores are relatively light, containing gas vacuoles and they are of various shapes. They are aerodynamically adapted for extended lateral travel through the atmosphere.

The passive liberation of spores into the atmosphere with air currents is common among microbes that produce dry spores on lateral mycelia (actinomycetes and fungi). Some spores are transmitted upward from microbial fruiting bodies by convection currents (surfaces exposed to sunlight have high temperature than the surrounding air and as they warm the air in contact with them, will set up convection currents which are of great importance in air movements). The effect of wind is to produce turbulence around stationary objects, though almost all surfaces are surrounded by a layer of still air, the laminar boundary layer (LBL), caused by the friction between the air and the object. It is essential for microbes to have some method of getting through this LBL if they are to land on or be dispersed from a surface. Others move laterally and vertically with the wind currents.

The higher the wind speed and lower the humidity, the greater the movement of spores. Many plant pathogenic fungi spread from one plant to another by this mechanism. Some spores are liberated only where water droplets in the air collide with the spore bearing bodies. For example, raindrops may liberate spores to the atmosphere. Spores and even vegetative microbes often enter the atmosphere as aerosols (splash from falling rain drops, spray from breaking waves, water striking rocks, sneezing and coughing) are important in the dispersal of some pathogenic bacteria and animal viruses.

In addition to the passive mechanisms that allow microbes to enter the atmosphere, there are some active mechanisms that discharge microbial spores into the atmosphere. In *Pilobolus*, the entire spore cluster is ejected when a vacuole in the sporangium base becomes turgid and thin and the air inside increases in osmotic pressure and then causes the structure to burst and the spores to be carried away in a jet of water to a distance of 1–2 m. In most Ascomycetes, ascospores are actively discharged.

Having become airborne, both spores and vegetative microbial cells face the problem of survival. Most organisms can survive a short passage (mm) through the atmosphere but relatively few survive long distance transport because dessication can cause microbes in the atmosphere to lose viability. During the day, microbes carried through the atmosphere on rafts such as soil particles or dust may be protected from the harmful effects of UV radiation.

Review Questions

1. Discuss air as an ecosystem.
2. Discuss atmosphere as a habitat for microflora.
3. Describe how spores are transported throught the atmosphere.

10

Sampling of Air

The goal of biological sampling is to help determine whether the biological particles present in a particular environment are affecting or causing irritation in certain individuals. Sampling is also used to locate the sources of indoor microorganisms and facilitate an effective remedy. While we are typically surrounded by a wide variety of microorganisms every day, sampling provides us with a method to establish in a scientific way whether the environment in question contains more organisms than would normally be present. There are numerous techniques that may be used to evaluate the level of indoor microorganisms. We believe, however, that scientific comparisons are only possible when measured volumes of air are sampled and when results of surveys are expressed in terms of volumetric measurements.

As microbiologists, we are more concerned with the biotic environment of air rather than the abiotic features. The biotic feature of air is the bio-aerosol which is very important in terms of pathogenesis and the infectious agent (human, animal and plant).

The aero-microbiological pathway describes the launching of bio-aerosols into the air, the subsequent transport via diffusion and dispersion of these particles, and finally their deposition.

LAUNCHING

Launching is the process whereby particles become suspended within earth's atmosphere. The launching of bio-aerosols is mainly from terrestrial and aquatic sources, with greater airborne concentrations or atmospheric loading being associated with terrestrial sources than with aquatic sources. Launching into the surface boundary layers can include diverse mechanisms such as:

- air turbulence created by the movement of humans
- animals and machines
- the generation, storage, treatment and disposal of waste material
- natural mechanical processes such as the action of water and wind on contaminated solid or liquid surfaces
- the release of fungal spores as a result of natural fungal life cycles

Transport or dispersion is the process by which the movement of air is transferred to airborne particles thus resulting in dissemination of airborne microbes over long distances. Most common type of transport phenomenon ranges from 10 minutes (*submicroscale transport*) for 100 m which is common within buildings. *Microscale transport* ranges from 10 minutes to 1 hour and from 100 m to 1 km and is the most common type of transport phenomenon. Microorganisms have limited ability to survive when suspended in atmosphere. However, viruses, spores (fungal) and spore forming bacteria can survive transportation up to 100 km in the atmosphere. For example, coliforms aerosolised from sewage treatment plants have been transported over 1.2 km. Bio-aerosols travel with the force of diffusion or dispersion. *Diffusion* is scattering of bio-aerosols in response to gravity and is aided by airflows and atmospheric turbulence.

TRANSPORT OF MICROORGANISMS IN AIR

The transport of bio-aerosols is primarily governed by hydrodynamic and kinetic factors while their fate is dependent on their specific chemical make-up and the meteorological parameters to which they are exposed. The vast majority of airborne microbes are immediately inactivated as a result of environmental stresses (dessication, temperature, and oxygen).

Bio-aerosol particles can be either solid or liquid and can come from a number of natural and anthropogenic sources.

Naturally occurring bio-aerosols are injected into the atmosphere either by chance (wind, rain, bursting bubbles, etc.) or by processes governed by natural selection.

DEPOSITION AND ADHESION

Deposition is in part governed by the mass of the particle. Larger particles often have the momentum to deviate from the streamlines, impact the surrounding surfaces and be deposited. Particle bounce also increases with increasing mass. Additional forces include gravitational settling, convection (due to temperature variations), diffusion and eddy diffusion.

Adhesive forces are dominated by the molecular structure and organisation near the contact surfaces. Two types of forces are primarily responsible for the at-a-distance interactions (within very short distances) that occur between aerosol particles and surfaces. If liquid is present, interfacial reaction (meniscus formation) can develop. Adhesion is also governed by the atomic attraction between surfaces and geometrical factors (shape of surfaces, flat versus spherical).

RELEASE

Adhesion can be overcome by the mechanism of aerodynamic forces and the release of biological particles on surfaces can be initiated. The mechanical energy frequently originates from turbulent mixing of the wind. The force required to remove particles increases with decreasing particle size. The surfaces of bio-aerosols are complex structures which include interactions between proteins, phospholipids, peptidoglycans, etc.

TYPES OF DEPOSITION

An airborne bio-aerosol will ultimately be deposited on a surface by one or more mechanisms like gravitational settling, surface impaction, rain deposition, electrostatic deposition, etc.

Gravitational Settling

The force of gravity acts upon all particles heavier than air, pulling them down and essentially providing spatial limitations on the spread of airborne particles. Gravitational settling is the function of the earth's gravitational pull, particle density, particle diameter and viscosity of air. Fig. 10.1 depicts the schematic representation of gravitational setting.

Downward molecular diffusion is a randomly occurring process caused by natural air currents that promote the downward movement of airborne particulates (in still as well as turbulent air). Molecular diffusion is also influenced by the force of wind.

Fig. 10.2 shows the schematic representation of downward molecular diffusion, a naturally occurring process caused by the air currents that promote and enhance gravitational settling of airborne particles. Although molecular diffusion can occur in any direction, due to the effects of gravity, the overall trend of the process results in net downward movement and deposition.

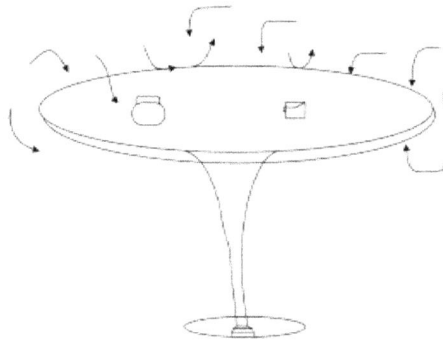

Figure 10.1 *Schematic representation
of gravitational settling*

Figure 10.2 *Schematic representation
of downward molecular diffusion of air*

Surface impaction

Particles make contact with surfaces such as leaves, trees, walls, etc. With impaction, there is an associated loss of kinetic energy. Impaction potential is the likelihood that an airborne object will collide with another object in the path leading to bouncing (rather than deposition). Bouncing causes the particle to re-enter the air when it can either allow gravitational settling to occur resulting in deposition or it can allow the particle to escape and enter the air current (Fig. 10.3). Impaction is influenced by velocity and size of the particle and size and shape of the surface it is approaching.

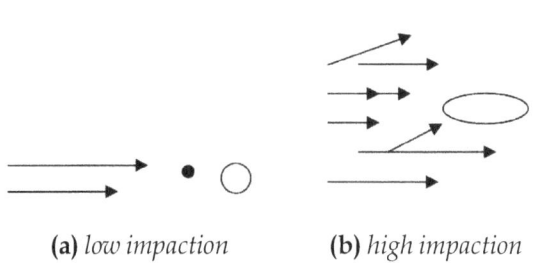

(a) *low impaction*   (b) *high impaction*

Figure 10.3   *Schematic representation of surface impaction*

Consider the two cases shown in Fig. 10.3. A small airborne particle travelling slowly toward another small round object (a) has much less chance of impaction with the small object, whereas a larger particle (b) travelling at a greater velocity towards a much larger flat object has a higher impaction potential.

Rain and Electrostatic Deposition

Rainfall deposition occurs as a condensation reaction between a raindrop and bio-aerosol which combine and create a bio-aerosol with a greater mass which settles faster. Electrostatic deposition also condenses bio-aerosol but is based on electrovalent particle attraction (on charges found outside the aerosol). All particles tend to have some type of associated charge. Microbes have an overall negative charge associated with their surface (at neutral pH). These negatively charged particles can associate with other positively charged aerosols resulting in electrostatic condensation thus increasing the bio-aerosol mass and enhancing deposition.

Having seen the basis of air sampling processes, we will now take a look at the methods of air sampling based on the above mentioned phenomena. Choosing an appropriate sampling device is based on availability, cost, volume of air to be sampled, motility, sampling efficiency and environmental conditions under which sampling will be conducted. Another factor which is important for air sampling is the microbial viability during and after sampling.

METHODS OF AIR SAMPLING

The various methods of air sampling include:

1. Impingement which is trapping of airborne particles in a liquid matrix.

2. Impaction which is forced deposition of airborne particles on a solid surface.

3. Centrifugation which is mechanically forced deposition of airborne particles using inertial forces of gravity.

4. Filtration which is trapping of airborne particles by size exclusion.

5. Deposition which is collection of airborne particles using only naturally occurring deposition process.

The two most commonly used devices for microbial air sampling are *all glass AGI-30 impinger* and *Anderson 6-stage impaction sampler.* Apart from these, the *Reuter air sampler* (based on centrifugation) is also commonly used for small-scale air sampling.

Impingement

An impinger (Fig. 10.4) operates by drawing air through an inlet that is similar in shape to the human nasal passage. The air is transmitted through a liquid medium where the air particles become associated with the fluid and are subsequently trapped. The AGI-30 impinger (All Glass Impinger) is a liquid-filled cylinder which collects particles by their impingement into a fluid. The capillary tip of the inlet tube inside the cylinder is located 30 mm from the impinger bottom, thus the nomenclature. AGI - 30 is easy to use, inexpensive, portable, reliable, easily sterilised and has high biological sampling efficiency in comparison to many other sampling devices but there may be loss in viability of the collected biological sample due to sheer force used to collect the air. The usual volume of collection medium is 20 ml and the typical sampling duration is approximately 20 minutes which prevents evaporation during the sampling of warm climates or freezing of the liquid medium when sampling at lower temperatures. The liquid and suspended microorganisms can be concentrated or diluted by using this method of impingement.

A simple medium is 0.85% sodium chloride which is an osmotically balanced, sampling medium used to prevent osmotic shock of recovered organisms. Another medium in use is peptone (1%) which is used as a medium for stressed organisms. Finally, enrichment medium can be used to sample selectively for certain types of organisms.

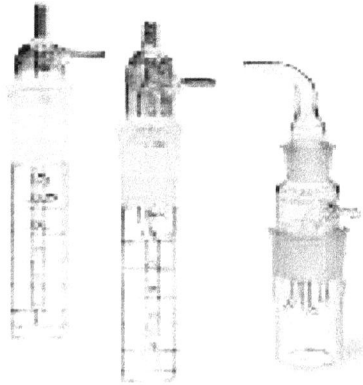

Figure 10.4 *AGI Impingers*

A major drawback when using an impinger is that there is no particle size discrimination which prevents accurate characterisation of the sizes of the airborne particles that are collected.

Impaction

The Anderson six-stage impaction sampler (Fig. 10.5) provides accurate particle size discrimination in contrast to the impinger. This device was developed by Anderson in 1958 and the general operating principle is that air is sucked through the sampling port and strikes agar plates. Impaction procedure depends upon the internal properties of the particle (size, density), on the physical parameters of the impactors (inlet nozzle dimensions) and the airflow pathway.

The principle that underlies this sampling device is simple and ingenious. Air impinging onto the top agar plate is travelling at relatively low speed, and is deflected around the agar plate. Only the larger (heavier) airborne particles will have sufficient momentum (defined as mass ´ velocity) to break free from this air current and impact onto the top agar surface. But then the same volume of air is sucked through a series of small holes, so its velocity is increased and this enables smaller particles to impact onto the second agar plate and so on, down the series of plates with increasingly smaller holes, so that the momentum of the airborne particles is increased at each stage.

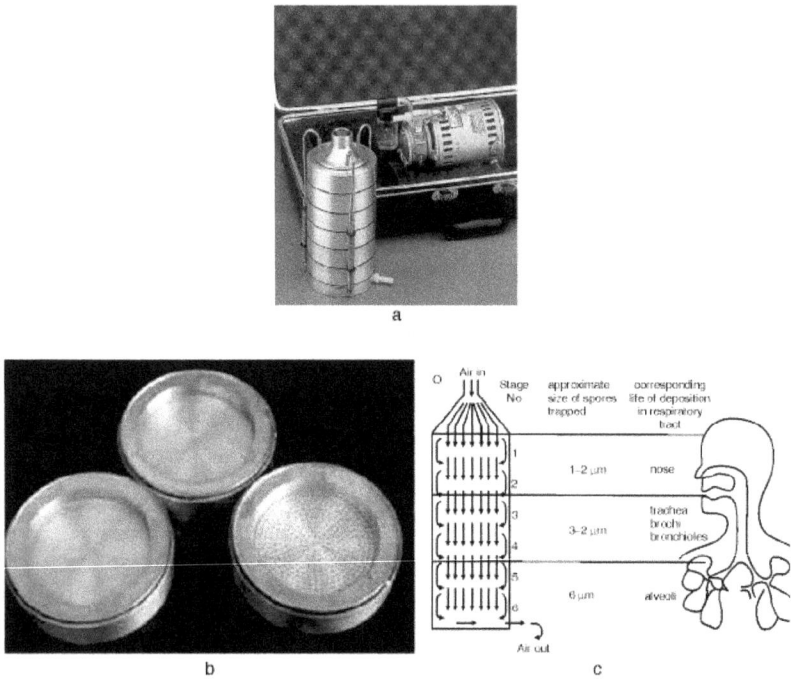

Figure 10.5 *The Anderson air sampler*

The result is a size (mass) separation of the airborne particles, which is remarkably similar to that which occurs in the human respiratory tract (Fig. 10.5c), as explained later. Large spores such as those that impact onto a rotorod tape (greater than 7–10 micrometres) will impact onto the topmost agar plate. Smaller spores (3–7 micrometres) will impact on the middle agar plates, and even very small spores (e.g. the spores of actinomycetes, 1–2 micrometres diameter) will impact on the lowest agar plates.

When the apparatus is running (see Fig. 10.5c) the incoming air impinges onto the topmost agar plate, where airborne particles can impact on the agar surface. Then the air is drawn round this first agar plate, and through the first set of perforations, so that particles can impact on the second agar plate, and so on down the stack.

Larger particles are collected on the first layer, and each successive stage collects smaller and smaller particles by increasing the flow velocity and consequently the impaction potential. The particle size distribution of the air particles can be directly related to the particle size distribution that occurs naturally in the lungs of animals.

The lower stages correspond to the alveoli and the upper stages to the upper respiratory tract. The biological sampling efficiency is somewhat lower because of the method of collection which is impaction on an agar surface (on a solid surface).

Figure 10.6 *Mechanisms of collection with impaction*

Fig. 10.6 depicts the mechanisms of collection utilised in bio-aerosol sampling with impaction.

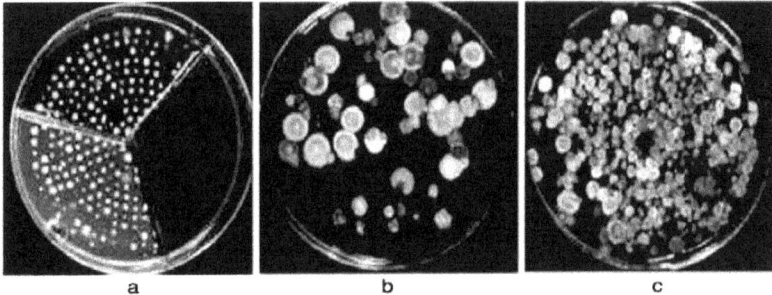

Figure 10.7 *Pictures showing the removed and incubated agar plates*

After this apparatus has run for some time (a few minutes or several hours, depending on the likely spore load in the air), the agar plates are removed and incubated, to identify the organisms that grow on them. Fig. 10.7a shows a plate from the lowest part of the Anderson sampler. The colonies are of thermophilic actinomycetes (*Faenia rectivirgula* or *Thermoactinomyces vulgaris*) that are common causes of Farmer's lung disease.

Figs. 10.7 (b) and (c) show agar plates from the middle part of the Anderson sampler, where several species of *Aspergillus* and *Penicillium* have developed from spores about 3–5 micrometres diameter.

Centrifugation

Centrifugal samplers (Fig. 10.8) use circular flow patterns to increase the gravitational pull within the sampling device in order to deposit particles. The most common is the Reuter's air sampler which is based on the centrifugal force. Air is sucked in through the propeller blades and as it traverses through the body of the sampler, it gets deposited onto the thin agar media which lines the inner wall of the sampler. After a known period of time the media is taken out and incubated for further studies.

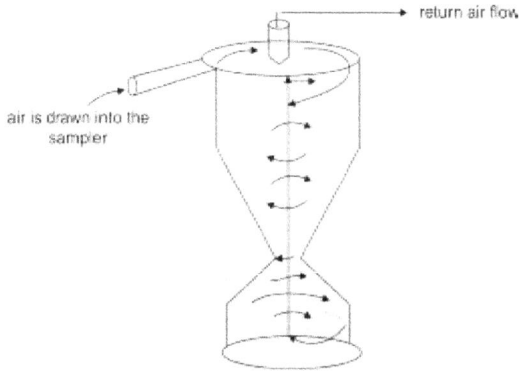

Figure 10.8 *Diagrammatic representation of centrifugal air sampler*

Figure 10.9 *SKC aluminum cyclone*

There is another device based on this principle (Fig. 10.9), the cyclone which is a tangential inlet and return flow sampling device. These samplers are able to sample a wide range of air volumes (1–400 L/min), depending on the size of the unit. The unit operates by applying suction to the outlet tube, which causes air to enter the upper chamber of the unit at an angle. The flow of air falls into a characteristic tangential flow pattern which effectively circulates air around and down along the inner surface of the conical glass housing. As a result of the increased centrifugal forces imposed on particles in the air stream, the particles are sedimented out. Analysis is performed by rinsing the sample with an appropriate liquid medium, collection of the medium and subsequent assay by standard methodologies.

Filtration and Deposition

Both the methods (Fig. 10.10) are widely used for microbial sampling for cost and portability reasons. Filter sampling requires a vacuum source and involves passage of air through a filter, where the particles are trapped. After collection, the filter is washed to remove the organisms before analysis. Usually membrane filters are used for the purpose with varied pore sizes.

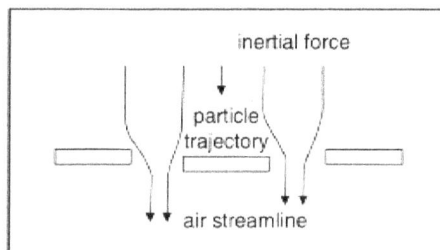

Figure 10.10 *Schematic representation of filtration methods*

Deposition sampling is by far the easiest and most cost-effective method of sampling. Deposition sampling can be accomplished merely by opening an agar plate and exposing it to the wind, which results in direct impaction, gravity settling, and other depositional forces.

There are problems associated with this method of sampling. They have low overall sampling efficiency because it relies on natural deposition. It has no defined sampling rates or particle sizing, and poses an intrinsic difficulty in testing for multiple microorganisms with varied growth conditions.

Analysis of microorganisms collected by depositional sampling is similar to impaction sample analysis.

There are other simple sampling methods that may be used to supplement volumetric air sampling. Surface samples are taken by tape lift imprint, by swabbing the suspect surface with a culture swab, or by submitting a bulk sample of the suspect surface.

For tape samples, a piece of absolutely clear (not frosted) tape that is one or two inches in length is used. It is handled by the ends only. The adhesive side of the tape is positioned over the suspect area and the tape is gently and firmly pressed. Care is taken that the tape is not rubbed back and forth. The tape is removed from the surface and placed on a clean microscope slide, which is put into a slide box or into plastic bags and further processed.

A direct microscopic examination is performed on surface samples. While culturing, a surface sample may help resolve a specific identification problem. Used alone, such a culture may result in an inaccurate characterisation of the surface sampled.

A direct microscopic examination of a surface shows exactly what is there, without any skewing by laboratory procedures. Surface sampling is inexpensive and (for a direct examination) may be analysed immediately. Surface sampling may also reveal indoor reservoirs of spores which have not yet become airborne.

The primary purpose of a direct microscopic examination of a surface is to determine whether or not mold is growing on the surface sampled, and if so, what kinds of molds are present. Secondarily, most surfaces collect a mix of spores which are normally present in the environment. At times it is possible to note a skewing of the normal distribution of spore types, and also to note 'marker' genera which may indicate indoor mold growth.

In addition, when mold growth is present indoors, many more spores of a particular type will be found trapped on surfaces. These spores may be in forms which indicate recent spore release (close proximity), such as spores in chains or clumps. Marker genera are those spore types which are present normally in very small numbers, but which multiply indoors when conditions are favourable for growth. These would include cellulose digestors such as *Chaetomium*, *Stachybotrys*, and *Torula*. While a single *Stachybotrys* spore is occasionally seen as part of the normal outdoor flora, finding 5 or 6 of these spores on a single scotch tape slide of a duct surface is an indicator that *Stachybotrys* may be growing indoors.

But the presence of biological materials on a particular surface is not a direct indication of what may be in the air. Health problems related to indoor microbial growth are generally caused by the inhalation of a substantial number of airborne spores, sometimes over a substantial period of time (exceptions being, for example, situations involving small children or immuno-compromised individuals).

The various sampling methods and their characteristics are summarised in Table 10.1.

Table 10.1 *Sampling methods and their characteristics*

Sampler type	Principle Flow rate(l min–1)	Cut-off 110 diameter(m50)	(/m)
Sieve Impactor (Anderson)Slit	impaction onto agar plate	28.3	0.65–7.0
Sampler (e.g. Casella)	impaction onto rotating agar plate	30–700	~0.5
Centrifugal Impactor(RCS)	impaction due to centrifugal acceleration	40	4.0
Impingers (e.g. AGI)	impingement into liquid	12.5	0.3
PBI SAS Sampler (single-stage impaction)	onto agar plate impaction	90/180	2.0
Settle plates	gravity	non-volumetric	N/A
Contact plates	surface sampling	non-volumetric	N/A

Review Questions

1. Describe the aeromicrobiological pathway.
2. What are bio-aerosols?
3. Discuss deposition and adhesion as part of aeromicrobiology.
4. Describe in detail the various deposition types.
5. Give a detailed account of methods of air sampling.

11

Potential Hazards in Laboratory Techniques

Infection of laboratory workers is a common problem while handling microorganisms and in most of these cases, finding out the source of infection is very difficult. The problem is very severe in clinical microbiological laboratories where the microorganisms being handled are mostly pathogenic, although other research and development laboratories experience a similar problem.

Infection may occur by breathing in (inhalation) of infected aerosol, droplet nuclei or infected dust particles.

Aerosol is a cloud of small droplets of liquid in air. Most aerosols consist predominantly of droplets smaller than 0.1 nm in diameter which dry within a second or so to become solid residues called droplet nuclei. These droplet nuclei are very small hence may remain airborne for up to several hours and may directly land on the upper respiratory or lower respiratory tract and initiate infection.

Invisible aerosols are generated by any action that breaks the continuity of the surface of a liquid. Some examples include:

- withdrawal of a loopful from a broth culture
- bursting of the film of culture in a loop
- vibration of a wire loop during an inoculation procedure.
- sputtering of a charged loop during flaming
- mixing of a suspension with a loop or a mixing equipment
- removal of a wet stopper, screw cap or cotton wool plug.

- expulsion of residual fluid from a pipette
- letting liquid fall in drops into a container instead of pouring it smoothly down a side
- vigorous shaking or high speed mixing of liquid
- dropping and breaking of culture plates or tubes
- centrifuging of full tubes or tubes with wet rim and breakages in the centrifuge

Infective dusts which are particles small enough to remain airborne for minutes or hours can be produced by spillage of cultures onto the skin, clothing or floor.

- After drying, the spillage residue is readily fragmented by minor movements into a fine dust and disseminated into the air.
- Clouds of infective dust can be released by the opening of containers of freeze dried cultures or by the withdrawal of cotton wool plugs that have dried after being wetted with culture fluid.
- Convectional and other air currents dispense droplet nuclei and dust particles widely within the laboratories. Particles more than 5 μm diameter are deposited in the nose and throat while smaller particles reach the bronchi and the lungs.

HAZARDOUS PROCEDURES IN THE LAB

Care should be taken while utilising the procedures given below:

Syringe and needle

- 'Needle stick' injury is common during use or disassembly.
- Aerosol may be liberated from a vibrating needle on withdrawal from a vein or a culture.
- Splashing or spraying may be caused by the forceful ejection of contents.
- Skin, clothing or bench may be contaminated by leakage from syringe.

Pipetting

- Infective material may be ingested during mouth pipetting.
- Careless handling of the pipette may result in aerosol formation while transferring inoculum (when the broth is dropped into the tubes).

Inoculating loop

- Vibration of an inoculation loop especially if longer than 4 cm can cause splashing and aerosol production.
- Flaming of a wet loop or cooling a hot loop in an agar plate, mixing a slide agglutination test or spreading a film may also produce aerosol.

Petridishes

- Water of condensation on the agar or in the lid may become contaminated and spill onto fingers and bench.

Shaking/mixing

- Shaking produces an aerosol even in a closed container and the aerosol may be released on opening the container.

Centrifugation

- Vibration can generate aerosol within the container used for centrifugation.
- Careless loading or unloading, breaking during centrifugation or premature opening after breakage can lead to dissemination of culture fluids.

Freeze drying

- Opening of sealed ampoules is hazardous if done in the open occupied room. It should be done only inside the safety cabinet since it can generate aerosols.

Stoppering tubes

- As previously discussed, opening tightly sealed cork stoppers, screw capped tubes or even cotton wool plugs can cause splashing, leading to aerosol formation.

Microbiological safety cabinets

- Unless properly installed, these cabinets can release infection into the air of the room or exhaust duct.

Animal procedures

- Working with animals like during inoculation, collection of samples or performing necropsy (autopsy) can account for lot of injury.
- Bedding contamination with excreta or urine can liberate infected dust.

Transport of specimen

- Improperly closed and packaged samples may leak onto the wrapping and contaminate the surroundings.
- Outside of the container may get contaminated during specimen collection.

Disposal of cultures

- Contamination may occur during decontamination of discarded cultures, specimen containers and used equipments (scalpels, needles).

Laboratory associated infections are quite serious and there is considerable rate of mortality. Safety in the lab is the responsibility of all those working in it.

Review Questions

1. What is aerosol?
2. How are invisible aerosols released into the atmosphere?
3. What are infective dusts? Give examples.
4. Detail out the various hazardous procedures in the lab.

12

Airborne Diseases

A mong various modes of disease transmission, air is one of the important routes and a number of diseases have been shown to be transmitted through air. Since human beings and animals are continuously inhaling the air, the chances for airborne microorganisms to find a host and cause infection are more. Most of the respiratory tract infections are acquired by inhaling the air containing the pathogen. Microorganisms in droplets and infectious dusts and spores can be easily disseminated through air.

Airborne infections cannot be defined as precisely as waterborne or food-borne infections, but they will be taken to include all those infections whose causative agents produce infection after inhalation. Thus airborne transmission refers to the spread of agents of infection by droplet nuclei in dust that travel more than 1 metre from the reservoir (source of infection) to the host. For example, the virus that causes measles and the tubercle bacilli can be transmitted via airborne droplets.

Broadly they can be of three kinds:

1. *Infection of the respiratory tract* like common cold, influenza, sore throat and pneumonia.

2. *Specific fevers* like measles, whooping cough, diphtheria, scarlet fever, etc.

3. *Specific respiratory infections*, the most important of which is pulmonary tuberculosis.

In addition, some infections may be acquired by inhalation of dried, infective excreta or secretions, e.g. psittacosis and Q fever.

MODES OF TRANSMISSION

There are three main methods of transmission of airborne infections:

Direct Droplet

It requires intimate contact and it may be difficult to be sure whether the infective material has been directly inhaled from the dispersed droplets or whether it has reached the respiratory tract of the susceptible person indirectly through the medium like hands, fomites etc. Direct droplet spread would seem to be the most likely method for transfer of respiratory pathogens that have a poor viability outside the body, e.g. cerebrospinal fever (associated with conditions of overcrowding in jails and ships). Another example is whooping cough. The infecting microbes are in the active phase of growth and therefore are more likely to be able to attack the tissues in which they are implanted.

Direct Airborne

Most of the droplets that are expelled from the mouth during talking, coughing and sneezing fall quickly to the ground or onto the clothing of the person in question. A proportion of the expelled droplets are however so small that by rapid evaporation they remain suspended in the air like particles of smoke and may be carried to considerable distances by air current from the coughing or sneezing individuals by air currents. These tiny droplets may contain infective material (virus infection) with the result that individuals at some distances from the infectious person may become infected. Among human infections, measles and chicken pox are spread in this way. It is inherent in this mode of spread that the infecting dose must often be very small.

Indirect Airborne

Most of the larger droplets expelled during talking and coughing become dried particles of dust on exposed surfaces or on clothing, bed cloths or handkerchieves. If they contain respiratory pathogens that resist natural drying, this dust when raised in the air during activities such as dusting, sweeping, bed-making or shaking of a handkerchief could be inhaled and may set up infection. Dust-borne spread has been proved for certain viral and rickettsial infections such as psittacosis and Q fever. Dust particles can harbour various pathogens. Staphylococci and streptococci can survive on dust and be transmitted by the airborne route. Spores produced by certain fungi are also transmitted by the airborne route and can cause such diseases as histoplasmosis, coccidioidomycosis and blastomycosis.

A number of potentially pathogenic microorganisms are part of the normal microbiota in the upper respiratory tract. They do not cause illness because the predominant microorganisms of the normal microbiota suppress their growth by competing with them for nutrients and producing inhibitory substances.

Airborne pathogens make their first contact with the body's mucous membranes as they enter the upper respiratory tract. Many respiratory and systemic diseases initiate infections here.

DISEASES OF THE UPPER RESPIRATORY SYSTEM

The upper respiratory tract consists of external nares and upper part of the pharynx.

Bacterial Diseases

Streptococcal pharyngitis (strep throat)

- It is an upper respiratory tract infection caused by group AB-haemolytic streptococci — *Streptococcus pyogenes.*
- The pathogen is resistant to phagocytosis.
- This organism can be detected in the throat of many people who are asymptomatic carriers.
- This is characterised by local inflammation and fever with tonsillitis.
- Another complication is *otitis media* (infection of the middle ear).
- Penicillin is the drug of choice.

Scarlet fever

- When the *Streptococcus pyogenes* strain produces an erythrogenic (reddening) toxin the infection is called scarlet fever.
- It is a communicable disease spread mainly by inhalation of infective droplets from an infected person.

Diphtheria

- Until 1935, it was the leading infectious killer of children.
- Disease begins with a sore throat and fever followed by general malaise and swelling of neck.

■ Causative organism is *Corynebacterium diphtheriae* which is a gram positive non-spore-forming rod.

■ The bacterium is well-suited to airborne transmissions and is very resistant to drying.

Otitis media

■ One of the uncomfortable complications of common cold.

■ Caused by *Streptococcus pneumoniae, Haemophilus influenzae, Moraxella catarrhalis, S. pyogenes* and *Staphylococcus aureus.*

Viral Diseases

Common cold

■ Caused by rhinoviruses (50%), coronaviruses (15–20%). 10% of all colds caused by several other viruses.

■ There are at least 113 serotypes of rhinoviruses.

■ Symptoms are sneezing, excessive nasal secretions and congestion.

■ Rhinoviruses thrive at a temperature slightly below that of normal body temperature (found in the upper respiratory tract).

■ A single rhinovirus particle deposited on the nasal mucosa is sufficient to cause a cold.

DISEASES OF THE LOWER RESPIRATORY TRACT

Lower respiratory tract consists of larynx, trachea, bronchial tubes and alveoli (air sacs that make up the lung tissue).

Bacterial Diseases

Pertussis (**whooping cough**, *per*-thoroughly and *tussis-cough*)

■ Causative organism is *Bordetella pertussis* which is a small obligately aerobic capsulated gram negative coccobacillus.

■ It is primarily a childhood disease and can be quite severe. Initial stage resembles a common cold.

■ Mortality rates among children is quite high.

■ Disease is transmitted by inhaling pathogens expelled by the coughing of the infected person.

■ Transmission rate is 90% among non-immune contacts.

Tuberculosis

- Causative organism, *Mycobacterium tuberculosis* is a slender acid-fast rod and obligate aerobe which is a slow grower.
- On the surface of the liquid media, their growth appears mold-like hence the name *Mycobacterium*.
- TB is most commonly acquired by inhaling the tubercle bacillus.

Bacterial pneumonias   Pneumonia is a pulmonary infection caused by bacteria.

Pneumococcal pneumonia

- Caused by *S. pneumoniae* which is a gram positive, capsulated ovoid bacterium in pairs.
- Involves bronchi and alveoli with high fever, breathing difficulty and chest pain.
- Sputum is rust colour because of blood coughed up from lungs.

H. influenzae pneumonia

- Caused by gram negative coccobacilli.
- Has similar symptoms as common cold.

Mycoplasmal pneumonia

- Causative organism is *Mycoplasma pneumoniae.*
- They are fastidious organisms which lack cell wall and show very small colonies on solid media.
- The disease is endemic and is the common cause of pneumonia in children.
- Transmitted by airborne droplets.
- Infects the upper respiratory tract initially with low fever, cough and headache followed by the lower respiratory tract infection.

Legionellosis

- Legionnaires' disease is caused by an aerobic gram negative rod *Legionella pneumophila.* (pneumo-lung, phila-loving).
- Characterised by high fever of 105°F, cough and general symptoms of pneumonia.
- Microbes can grow in the water of air conditioners and cooling towers, the cause of epidemic in hotels and hospitals through airborne transmission.

Psittacosis (ornithosis)

- Term is derived from the diseases associated with psittacine birds (parakeets, parrots).

- Causative agent is *Chlamydia psittaci* which is a gram negative obligate intracellular bacterium.

- They produce tiny elementary bodies as part of their life cycle. These elementary bodies are resistant to environmental conditions, and therefore are transmitted through air.

- Psittacosis is a form of pneumonia that causes fevers, headache and chills with delirium (disorientation).

- One of the most common modes of transmission is inhalation of dried particles from droppings of birds.

Chlamydial pneumonia

- Causative organism is *Chlamydia pneumoniae.*

- Shows similar symptoms as pneumonia.

Q fever (query fever)

- Caused by a rickettsial member *Coxiella burnettii.*

- Symptoms are undulating fever (1–2 weeks) with chills, chest pain severe headache.

- This organism is resistant enough to survive airborne transmission.

- This pathogen is a parasite among arthropods (ticks) which transmit the organism to cattle and dairy herds following which the microbes are shed in milk, faeces and urine of infected cattle.

- Once the disease is established in a herd, it is maintained by aerosol transmission.

- The disease can be spread by inhaling aerosols of microbes generated in dairy barns, especially from placental material at calving time.

- Inhaling a single pathogen is enough to cause infection.

Viral Diseases

Viral pneumonia This occurs as a complication of influenza, measles or even chicken pox.

Respiratory syncytial virus (RSV)

- Most common cause of viral respiratory disease in infants.
- Symptoms are coughing and wheezing lasting for more than a week.
- All children become infected by the age of two.

Influenza (viral flu)

- Characterised by chills, fever, headache and muscular ache.
- Caused by influenza virus.

Fungal Diseases

Fungi often produce spores that are airborne and are causative agents of various diseases.

Histoplasmosis

- It resembles TB superficially.
- Symptoms are sub-clinical.
- Caused by *Histoplasma capsulatum* which is a dimorphic fungus (yeast- like inside the tissue and filamentous in artificial media).
- Disease is acquired from airborne conidia produced under conditions of appropriate moisture and pH.

Coccidioidomycosis

- Causative organism is *Coccidioides immitis* which is a dimorphic fungus. Within tissues, the organism forms a thick-walled body filled with spores called spherules. In soil, it forms filaments which reproduce by formation of arthrospores which are carried by wind to transmit the infection.
- Laboratory workers must take great care because of the possibility of infectious aerosols.

Pneumocystis pneumonia

- Causative organism is *Pneumocystis carinii*.
- It has characters of both a fungus and a protozoan but it is strongly related to certain yeasts (RNA analysis).
- Causes endemics in hospitals.
- The pathogen is found in healthy human beings but causes disease among the immuno-compromised persons.

- People with AIDS are very susceptible to this pathogen.
- This organism is found in the alveoli of the human lungs.
- Symptoms of the infection resemble that of pneumonia.

Blastomycosis

- Causative organism is *Blastomyces dermatitidis* which is a dimorphic fungus.
- Infection begins in the lungs and spreads rapidly.

Aspergillosis

- Causative organism is *Aspergillus fumigatus*.
- Common among gardeners and farmers due to the spores which are widespread in decaying vegetation.

Rhizopus and *Mucor* species also cause pulmonary infections.

Review Questions

1. Define airborne infections.
2. What are the types of airborne infections?
3. Discuss modes of transmission of various airborne infections.
4. Give an account of bacterial airborne diseases.
5. Give an account of fungal airborne diseases.

13

Air Sanitation

A ir sanitation is the system of removing the impurities present in air inside buildings to protect people from infections. Sanitation of air is essential in enclosed places like hospital wards, operation theatres and burns unit to prevent infection. Food processing and packaging industries and rooms where sterile materials or products are stored require aseptic atmosphere and safe handling to prevent contamination.

The factors that determine how conducive a particular building is to spreading disease include the following:

- The range of temperature and humidity control
- The amount and distribution of outdoor air
- The efficiency of the filters
- The cleanliness of the facility
- The number and types of surfaces throughout the building
- The hygiene of the occupants

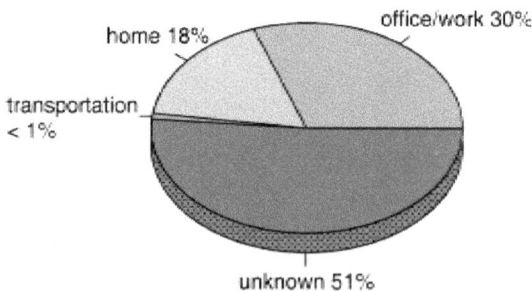

Figure 13.1 *Sources of infection in adults*

We are well-aware of the laboratory hazards with respect to the bio-aerosol formation thus leading to airborne infections, with the laboratory personnel more prone to it. With proper installation and maintenance of air sanitation devices within a lab, we can definitely minimise such airborne infections and guarantee the safety of the laboratory personnel.

Many labs work with pathogenic microorganisms, some of which are dangerous particularly those associated with the air.

The various devices used for air sanitation include the following:

Ventilation

Ventilator is an opening or a device that allows fresh air to enter into an enclosed space. A great reduction in the airborne microorganisms can be brought about by laminar airflow system developed by Whitfield. It was found that a horizontal flow of air at about 40 feet/minute in a room in which spores of *Bacillus subtilis* had been liberated, led to a diminution in their numbers by 100–1000 times greater than that to be expected in a room where turbulent ventilation was employed. The ultimate criterion of its effectiveness must be the diminution of cross infection.

Safety Cabinets

They are essentially isolation chambers that provide a safe environment for the manipulation of pathogenic microbes. They are characterised by having

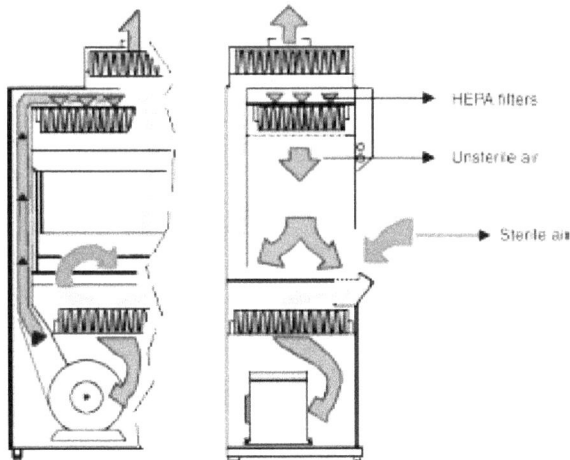

Figure 13.2   *Class I cabinets*

considerable negative pressure air flow that provides protection from infectious bio-aerosols generated within the cabinets. They are of three classes.

Class I Exhaust protection cabinets They protect the operator by maintaining a rapidly moving stream of air, which entrains infectious particles and deposits them on a high efficiency filter. The air is then discharged outside the building (Fig. 13.2)

Class II Laminar flow cabinets These are open-fronted and are designed to protect the material that is being handled from extraneous contamination. It is suitable for most work with moderate pathogens. They do not protect the operator, though they can be modified to do so (Fig. 13.3).

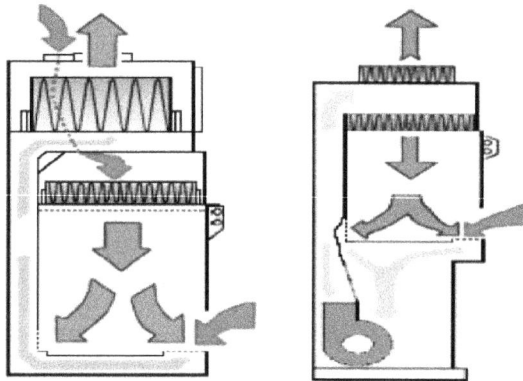

Figure 13.3 *Class II cabinets*

Class III cabinets These are used for the handling of very dangerous materials and are completely enclosed. Materials are handled through gas-tight glove ports and air enters through a non-return valve and is extracted through a high efficiency filter (Fig. 13.4).

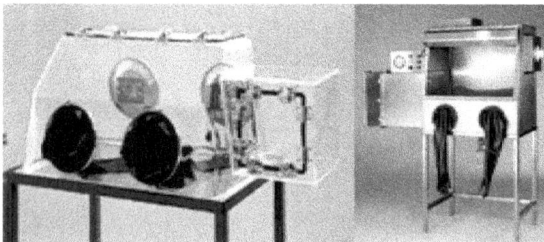

Figure 13.4 *Class III cabinets*

Disinfectant Sprays and Vapours

Earlier, carbolic spray was introduced by Lister. Sulphurous acid and formalin vapours are occasionally still used for terminal disinfection in hospitals but they are highly irritating to the respiratory passage. Finding by Douglas *et al* was very promising where it was possible to destroy the cells of *E.coli* in the atmosphere by means of sodium hypochlorite solution sprayed into the air in very high dilution. After a lot of research, spraying of 1% sodium hypochlorite solution at a concentration of 1–5/million parts in air was agreed upon. Other materials of spray include, resorcinol, propylene glycol, triethylene glycol and ozone but none of them have proved their worth. Ozone and 2.5% of hydrogen peroxide play a very important role in air disinfection since they can kill cells of *Salmonella typhi* most effectively. To both UV light and disinfectant vapours, gram-negative bacilli are more susceptible than gram positive organisms. Microorganisms are harder to kill when dried onto dust particles than when they are in the form of droplet nuclei. The different techniques for air sanitation are shown in Table 13.1.

Table 13.1 *Different techniques for air sanitation*

Airborne Pathogen Control Technologies	
Current	**Developmental**
Isolation Systems	Photocatalytic Oxidation
Air Filtration	Air Ozonation
Ultraviolet Irradiation	Carbon Adsorption
Outdoor Air Purging	Passive Solar Exposure
Electrostatic Precipitation	Ultrasonic Atomisation
Negative Air Ionisation	Microwave Atomisation
Vegetation	Pulsed Light

CURRENT AIR DISINFECTION METHODS

There are various technologies currently in practice to control airborne pathogens. The current air disinfection methods are listed below.

Isolation Systems

These can be classified into three basic categories:

    a.   Negative pressure isolation rooms

b.   Positive pressure isolation rooms

c.   Multi-level biohazard laboratories

Filtration of Microorganisms

Three types of filters exist for use in ventilation systems, pre-filters, HEPA (High Efficiency Particulate Air) filters and ULPA filters (Fig. 13.5 a,b).

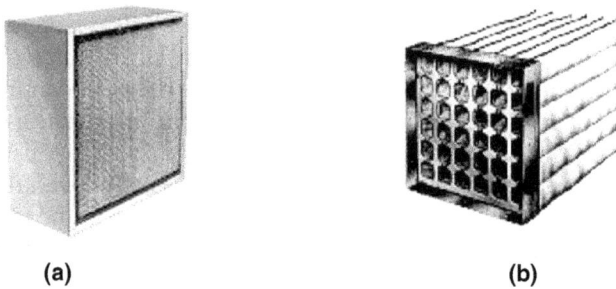

(a)   (b)

Figure 13.5   *Filters for air sanitation*

A typical HEPA filter, such as the one shown in Fig. 13.5b will filter micron sized particles at about 95% efficiency. Some box or pleated type filters (Fig. 13.5a) can be as thin as 2–4 inches, or as wide as 8–12 inches. The picture at the right shows a bag type HEPA filter, which can extend upto 24 inches. Bag type filters typically have a lower pressure drop than the pleated or box type HEPA and are more efficient.

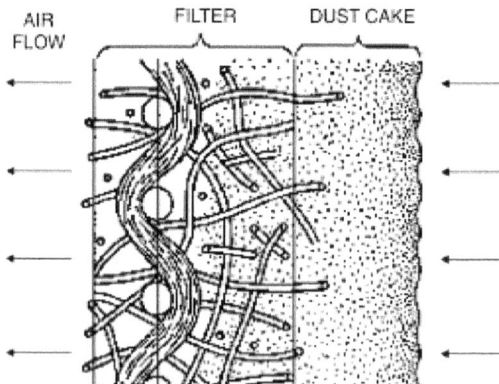

Figure 13.6   *Illustration of a HEPA filter*

HEPA filters are commonly found in hospital isolation rooms, operating theatres, and Level 3 and 4 containment facilities, as well as in industrial clean rooms.

HEPA filters are typically rated as 99.97% effective in removing dust and particulate matter above 0.3 micron in size, based on DOP (diocytl phthalate) testing usually performed by the manufacturer. In theory, HEPA filters should be highly effective against bacteria and fairly effective against viruses, but real world installations do not always achieve performance limits measured in laboratories.

HEPA filters consist of fine fibres as illustrated in Fig. 13.6. Materials vary, but generally these are made of synthetic fibrous materials. The principle of HEPA filtration is not to restrict the passage of particulate by the gap between fibres, but by altering the airflow streamlines. The airflow will slip around the fibre, but any higher-density bio-aerosols or particulate matter will not change direction so rapidly and, as a result of their inertia, will tend to impact the fibre. Once attached, most particulates will not be re-entrained in the air-stream.

Ultraviolet Germicidal Irradiation (UVGI)

Microbes are uniquely vulnerable to the effects of light at wavelengths at or near 2537A°(Fig. 13.7) due to the resonance of this wavelength with molecular structures. Looking at it another way, a quanta of energy of ultraviolet light possesses just the right amount of energy to break organic molecular bonds. This bond breakage translates into cellular or genetic damage for microorganisms. The same damage occurs to humans, but is limited to the skin and eyes.

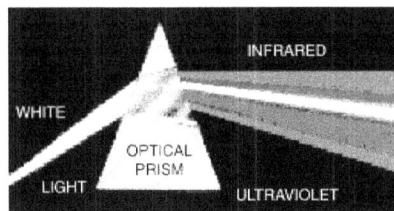

Figure 13.7 *Ultraviolet radiations*

The factors that can alter the effectiveness of UVGI, include the following:

■ Exposure time (the air velocity must allow for a sufficient dose).

- Room air mixing (for non-powered applications like ceiling units).
- Power levels.
- The presence of moisture or particulates providing protection for microbes.
- Dust settling on light bulbs can reduce exposures and thus maintenance is necessary.

Viruses are especially susceptible to UVGI, more so than bacteria, but are also very difficult to filter.

A combination of filtration for bacteria and spores, with UVGI for viruses may be an optimum combination, if all components are sized appropriately.

Outdoor Purge Air Systems

Airborne pathogens can be removed by purging with outside air, which is naturally sterilised. Airborne bacteria and viruses pathogenic for humans rarely occur in the outdoor air, and cannot survive long if they do. Spores of fungi and actinomycetes can occur in outside air but rarely occur in hazardous concentrations.

Electrostatic Precipitation

Electrostatic precipitators are commonly used to remove particles from airstreams having large steady flow rates. Typical applications include coal-burning plants and cement kilns. A typical two-stage electrostatic precipitator

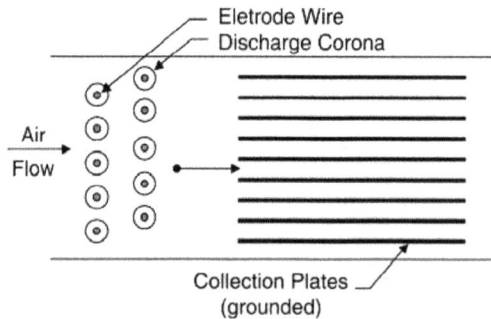

Figure 13.8   *Schematic representation of electrostatic precipitation*

has a stage of corona wires and a stage of collecting plates, as illustrated in Fig. 13.8. The corona wires are maintained at several thousand volts which produces a corona that releases electrons into the air stream. These electrons attach to dust particles and give them a net negative charge. The collecting plates are ground and attract the charged dust particles. The collecting plates are periodically rapped by mechanical rappers to dislodge the collected dust, which then drop into hoppers below. The air velocity between the plates needs to be sufficiently low to allow the dust to fall and not to be re-entrained.

Negative Air Ionisation

Ions are natural particles floating in the air around us all the time and have either a positive or negative charge. Scientific studies have shown that atmospheres charged with negative ions relieve hay-fever and asthma symptoms, seasonal depression, fatigue and headaches. A room charged with negative ions was shown to stem bacteria growth and precipitate many airborne contaminants including pollen, dust and dust mites, viruses, second-hand cigarette smoke, animal dander, odours and toxic chemicals. Negative air ionisation has the potential to reduce the concentration of airborne microorganisms. The effect appears to result from the ionisation of bio-aerosols and dust particles that may carry microorganisms, causing them to settle out more rapidly. Settling tends to occur on horizontal surfaces, especially metallic surfaces, and generally in the area near the ionisation unit. Ionisation may enhance agglomeration, creating larger particles out of smaller particles, and thereby increasing the settling rate. Ionisation may also cause attraction between ionised particles and ground surfaces. In situations where dust may carry microorganisms, negative air ionisation can be economical to use to reduce infections. Fortunately through modern technology it is possible to control the electrical state of our indoor environments by generating negative ions back in the air. These negative ions attach themselves to airborne toxins or to particles such as dust, pollen, smoke and dander and drop them from the air. Airborne microbial levels were reduced by 32–52% with ionisation in one particular study. It was also found that horizontal plates picked up considerably more cultures than vertical plates, strongly suggesting that settling out of ionised particles was the primary mode of removal.

Vegetation and Air Disinfection

A handful of studies have investigated the use of vegetation as a means of removing or reducing levels of airborne microorganisms. This has sometimes been referred to as 'growing clean air.' Recently a Canadian firm developed what it called a 'breathing wall,' or a wall of plants and waterfalls that seems to improve air quality.

The reasons that vegetation may reduce levels of airborne microorganisms are varied. The surface area of large amounts of vegetation may absorb or adsorb microbes or dust. The oxygen generation of the plants may have an oxidative effect on microbes. The increased humidity may have an effect on reducing some microbial species although it may favour others. The presence of symbiotic microbes such as *Streptomyces* may cause some disinfection of the air. Natural plant defences against bacteria may operate against mammalian pathogens.

CURRENT DEVELOPMENTS IN BIO-AEROSOL DISINFECTION

The methods described above are in common use. There are certain advanced techniques practised by the developed countries. These current techniques in air disinfection are as follows:

Photocatalytic Oxidation (PCO)

Titanium dioxide (TiO_2) is a semiconductor photocatalyst. When this material is irradiated with photons of less than 385 nm, an electron is promoted from the valence band to the conduction band. The resultant electron-hole pair has a lifetime in the space-charge region that enables its participation in chemical reactions.

Hydroxyl radicals and super-oxide ions are highly reactive species that will oxidise volatile organic compounds (VOCs) adsorbed on the catalyst surface. They will also kill and decompose adsorbed bio-aerosols. The process is referred to as heterogeneous photocatalysis or, more specifically, photocatalytic oxidation (PCO).

Several attributes of PCO make it a strong candidate for indoor air quality (IAQ) applications. Pollutants, particularly VOCs, are preferentially adsorbed on the surface and oxidised to (primarily) carbon dioxide (CO_2). Thus, rather than simply changing the phase and concentrating the contaminant, the absolute toxicity of the treated airstream is reduced, allowing the photocatalytic reactor to operate as a self-cleaning filter relative to organic material on the catalyst surface.

Photocatalytic reactors may be integrated into new and existing heating, ventilation, and air conditioning (HVAC) systems due to their modular design, room temperature operation, and negligible pressure drop. PCO reactors also feature low power consumption, potentially long service life, and low maintenance requirements. These attributes contribute to the potential of PCO technology to be an effective process for removing and destroying low level pollutants in indoor air, including bacteria, viruses and fungi.

Air Ozonisation

As shown in Fig. 13.9, ozone is injected into the airsteam and mixed in the turbulator to a degree that would guarantee ozonisation of all organic compounds, including viral nucleic acids and bacteria. Due to the corrosiveness of the ozone, an efficient reclamation system must be developed. Reclaimed ozone could be recycled to the injector, or else neutralised and used to regenerate electricity which would feed back to the regenerator.

Ozonisation has proven extremely effective in water systems, but as yet no air-side systems have been developed and proven safe and effective. Ongoing research at Penn State has found airborne concentrations of ozone highly effective in disinfecting surfaces.

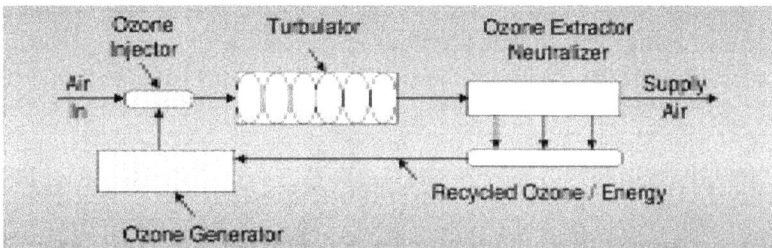

Figure 13.9 *Schematic representation of air ozonisation*

Carbon Adsorption

Carbon adsorption is used primarily for removal of gases and vapours. It is effective against volatile organic compounds (VOCs) but is not used for control of airborne dust or microorganisms. It is, in fact, not advisable to use carbon adsorption where particulate matter is present and may clog the adsorbent bed.

Carbon adsorption depends on the use of materials like activated charcoal which possess an enormous amount of surface area per unit mass. The presence of this surface area allows gas molecules to adhere to the surface.

Though carbon adsorbers are unlikely to have a significant effect on airborne microbes, they can be effective at removing VOCs generated by fungi and bacteria, and so decrease the health threats. Although it is not used for intercepting particulate matter, the use of carbon adsorption for the control of airborne viruses, which are not much larger than VOCs, is a potential application which remains to be studied. A mere ten-fold increase in pore size might be sufficient to adsorb viruses.

Ultrasonic Atomisation

Ultrasonics are capable of atomising water droplets, and in theory could atomise bacteria, which contain, or are contained in water. Viruses, which are either contained within droplets of water or have organic components such as DNA, RNA or proteins, should also be atomisable.

Two methods by which this may be accomplished are (a) supersonic nozzles and (b) sonic generators.

a. If the airstream is forced through a supersonic nozzle, a standing shock wave develops at the nozzle outlet. This shock wave dissipates energy by imparting it to the airstream, causing it to expand suddenly and rapidly. This results in the atomisation, or reduction to gas, of all bio-aerosols in the airstream. The fan power or pumping power required to accomplish this however, would be considerable.

b. The sonic generator (Fig. 13.10) essentially a high-power speaker and amplifier, tuned to resonate within the ductwork cavity, would create a standing shock wave through which the airstream would pass, and in which atomisation of any bio-aerosol would occur. Both this and the supersonic nozzle system would require a sound insulated ductwork section with inlet and outlet silencers.

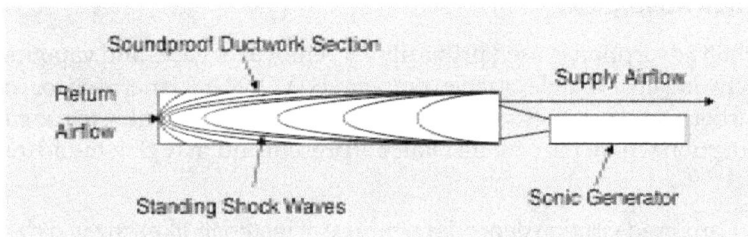

Figure 13.10 *Schematic representation of ultrasonic atomisation*

Microwave Atomisation

Microwaves consist of mutually perpendicular electrical waves and magnetic waves. Each of these components has an effect on the water molecules and other organic molecules which make up the bacterial cell or viral structure. The water molecules will rotate at or near the microwave frequency, and this energy translates into linear motion. Linear motion of gas or liquid liberates heat, and this thermal activity ultimately disrupts the cell and viral structures. Microwaves have been demonstrated to have biocidal effects due to the heating they induce, and are used to sterilise equipment. Normally this requires extended exposure time, but with a boost in power the exposure time could theoretically be reduced. In addition, there exists a phenomenon called the microwave effect which appears to destroy viruses for reasons other than heating.

Pulsed White Light (PWL) and Pulsed Electric Field (PEF)

Pulsed White Light (PWL), also called Pulsed Light or Pulsed UV Light, involves the pulsing of a high-power xenon lamp for about 0.1–3 milliseconds as per some sources or about 100 microseconds to 10 milliseconds as per other sources. The spectrum of light produced resembles the spectrum of sunlight but is momentarily 20,000 times as intense.

These high intensity flashes of broad spectrum white light pulsed several times a second can inactivate microbes with remarkable rapidity and effectiveness. The germicidal effect appears to be due to both the high ultraviolet content and the brief heating effects.

This technology is currently being applied in the pharmaceutical packaging industry where translucent, aseptically manufactured bottles and containers are sterilised in a once-through light treatment chamber. The chamber generates a light intensity at the surface of the exposed containers of about 1.7 J/sq.cm., or 1.7 × E06 microWatts/sq.cm. Sunlight produces about 1359 Watts/sq.cm.

Only two or three pulses are sufficient to completely eradicate bacteria and fungal spores.

The exact mechanism by which PWL kills bacteria and spores appears to be due to the effects of UV combined with a new disinfection mechanism¾ disintegration of the cell wall.

While UV causes damage to the nucleic acid and other components of the cell, the instantaneous heating of the cell results in the rupture of the cell wall, or lysing. This disintegrating effect has been demonstrated to occur in the absence of UV.

Pulsed Electric Field (PEF) involves the pulsing of electric fields of about 4–14 kV/cm through a liquid medium. The result of this momentary field is a membrane potential across the bacterial cell wall of more than 1.0 V, which is sufficient to lyse or damage the cell irreparably. The inactivation of various microbes, including *Escherichia coli, Lactobacillus brevis, Pseudomonas fluorescens, Bacillus cereus* spores, and *S. cerevisiae* has been found to be dependent on field strength and treatment times that are unique to each species. Since this method has little effect on proteins, enzymes, or vitamins, it is perfectly suited for food processing where the liquid medium may be anything from bouillon soup to milk. Only two or three pulses are sufficient to completely eradicate bacteria and fungal spores.

The exact mechanism by which PWL kills bacteria and spores appears to be due to the effects of UV combined with a new disinfection mechanism¾disintegration of the cell wall. While UV causes damage to the nucleic acid and other components of the cell, the instantaneous heating of the cell results in the rupture of the cell wall, or lysing. This disintegrating effect has been demonstrated to occur in the absence of UV.

Review Questions

1. Define air sanitation.
2. Discuss the role of safety cabinets in maintaining air sanitation.
3. Discuss in detail the current air disinfection methods.
4. Brief out the various developing air sanitation methods.

14

Air Pollution

Air emission from various industries or waste treatment processes release into the atmosphere, substances called pollutants that may be noxious or hazardous to humans. This excessive concentration of foreign matter in the atmosphere is termed as air pollution.

POLLUTANTS

The common pollutants are classified as particulate pollutants and gaseous pollutants.

Particulate pollutants

    a.   SPM (suspended particulate matter)

    b.   Bioparticles (organisms, spores, pollen)

    c.   Dust particles

    d.   Smoke

    e.   Mist

    f.   Fumes

    g.   Spray

    h.   Benzopyrenes

    i.   Asbestos

    j.   Pesticides

    k.   Metallic dust (arsenic, barium, boron, selenium, beryllium, cadmium, etc.)

Gaseous pollutants

a. Inorganic: NO, NO_2, SO_2, CO, CO_2, O_3, PAN, HF, NH_3, Cl, H_2S
b. Organic: hydrocarbons (CH_3, C_2H_4, C_2H_2, C_2H_5), aldehydes, alcohols

COMMON AIR POLLUTANTS

Sulphur Dioxide

It is one of the major air pollutants. It is emitted from burning fossil fuels like coal and petrol and from processing of sulphide ores like pyrite. When directly absorbed, SO_2 affects the morphology and physiology, and indirectly it leads to the formation of acids like sulphuric and nitric acid which are responsible for the low pH rain water. Spores of *Alternaria tenuis, Fusarium moniliforme* are tolerant to SO_2 whereas spores of *Rhizopus nigricans* and *Penicillium* species are comparatively susceptible. It is estimated that SO_2 from human activity introduced 6.6 million tons of sulphur into the atmosphere annually. Biologically produed H_2S, when oxidised, produces SO_2.

Effects

- SO_2 can damage materials and properties mainly through its conversion into the highly reactive H_2SO_4. It causes discolouration and physical deterioration of building materials and sculptures. Deterioration and fading are also produced in fabrics like cotton, nylon, leather and paper.

- It accelerates corrosion of metals, especially iron, steel and zinc.

- SO_2 has been found to affect vegetation adversely even at the concentration below 0.032 ppm. High concentrations of SO_2 over short period of time can produce acute leaf injury, such as necrosis in plants or brownish colouration in the tips of pine needles. Lower concentrations over long period leads to chronic leaf injury such as gradual chlorosis.

- SO_2 and H_2SO_4 both are capable of causing irritation in the respiratory tracts of animals and humans and high concentrations of SO_2 cause severe heart and lung diseases.

- SO_2 causes chlorosis and necrosis of vegetation in as low concentration as 0.032ppm. Lichen vegetation is completely destroyed.

- SO_2 pollution is related to higher death rate in aged persons. It kills fish and other animal life. Lichens and some bryophytes are found to be extremely sensitive to SO_2 pollution. The algal symbiont of a lichen thallus is most vulnerable to SO_2 pollution. They are regarded as indicators of SO_2 pollution.

Ozone

Ozone belongs to the class of secondary pollutants. Another secondary pollutant PAN (peroxyacylnitrates). Secondary pollutants are formed by the action of sunlight on the primary pollutants (directly released from the pollution sources).

Effects

- At a concentration of only 0.02 ppm, ozone destroys chlorenchyma cells and produces necrotic areas. There is a positive correlation between ozone concentration and destruction of foliage in a large number of plants.
- Ozone hardens rubber.
- It discolours and damages textiles.
- At less than 1 ppm, ozone injures mucous membrane.

Fluorides

Fluoride is an element which is present in every plant in small quantity. Even a low concentration of fluoride is effective in bringing about change in plant metabolism. Fluorides come next to SO_2 in the hierarchy of atmospheric gaseous pollutants. Fluorides are released in air as gaseous hydrofluoride and volatile fluorides like Na_2AlF_6 and SiF_4. These pollutants result from aluminium factories, brick kilns, pottery industries, ferro-enamel works and combustion of coal. Hydrogen fluoride is injurious at a concentration of 0.001 ppm to 0.10 ppm. Fluorides are accumulative poisons.

Carbon Monoxide

It is a colourless, tasteless and odourless gas which is deadly poisonous in higher concentrations. It is one of the important gaseous air pollutants. Human activities result in the production of nearly 250 million tons of CO annually. Aircrafts are responsible for about 2.5% CO. The toxic effects of CO on human beings or animals are the result of its reversible combination with haemoglobin in the blood. Haemoglobin has much greater affinity with CO and it lessens the oxygen carrying capacity of blood. It also reduces the dissociation of oxyhaemoglobin. Carbon monoxide accounts for 50% of the total atmospheric pollutants. It is formed by incomplete combustion of carbon fuels in various industries, motor vehicles, hearths, etc. A good amount of CO is also produced naturally by plants and animals. At 100 ppm concentration, CO produces giddiness and headache within less than an hour. Such a concentration is quite common at the time of traffic jams on the busy roads of a city. At higher concentrations, CO proves lethal.

Nitrogen Oxides

Biological production of NO and NO_2 amounts to about 1 billion metric tons annually while man's combustion processes produce 48 million metric tons of NO_2 annually. These oxides play an important role in the production of photochemical smog.

Methyl Isocyanate Pollution

The very term methyl isocyanate reminds one of the Bhopal air pollution tragedy of Dec 2, 1984. Methyl isocynate is an ester of isocyanic acid used in the manufacture of pesticides. First carbon monoxide is prepared by partial oxidation of coal or petroleum. CO is then combined with chlorine gas in the presence of activated carbon at high temperature to form phosgene. Phosgene is a deadly poisonous gas. This gas is passed in methylamine solution and methyl isocyanate is produced. It is used in the manufacture of carbonate insecticides. It kills insects at low doses but mammals are not affected so severely. In man, it causes burning sensation in eye, expels oxygen out of lungs causing death due to choking.

Particulate Matter

These are of two types: settleable and suspended. The settleable dust particle has a size larger than 10 μm. The smaller particles are able to remain suspended for long periods in air. They are further distinguished into two types: (i) less than 1μm aerosol (ii) more than 1 μm dust (if solid) and mist (if liquid). Heat produced in various industries and scattered by concrete, asphalt, etc. during daytime, helps in the dispersion of particulate matter.

Carcinogenic Pollutants

Polycyclic hydrocarbons (organic compounds having several fused rings), cigarette smoke condense and various chemical fractions of it are quite mutagenic. These substances are also implicated in cancer of the larynx, lung and gastrointestinal tract, and cause malignant transformation in tissue culture.

Asbestos

Asbestos is widely used as an insulating material and the chances of scrap asbestos reaching water bodies are quite appreciable. Asbestos occurs in four forms, namely chrysolite, crocidolite, remolite and amosite. The size of asbestos fibres in most of water supplies is very small (less than $0-1$ μm in diameter) hence it remains undetected by light microscopy. It causes cancer of lungs, stomach and small intestine. The dirt and particles which are carried from land to water constitute a very serious pollutant. A rare form of cancer, mesothelioma, developing in the skin membranes, lining chest and abdominal cavities, is strongly correlated with exposure to asbestos. About one-half of asbestos workers die of cancer, and the risks extend to the workers, wives and families who are exposed to asbestos dust and fibres on workclothes. The general public is exposed to asbestos fibres in the air from brake drums, insulation and the like and the food and water from diverse sources.

Arsenic

Long known to be poisonous, arsenic has now been shown to be almost certainly involved in producing both respiratory and skin cancers in human beings.

Nitrosamines

Nitrites are widely distributed in the environment and are commonly used in curing meat and fish. Nitrates are also common in soils and plants where microorganisms capable of reducing nitrates to nitrites are present. Secondary amines may be present in a variety of foods including cereals, mushrooms various vegetables and fruits, fish, cheese, tea and mince. They are also found in some drugs. These amines have been shown to react with nitrite to form nitrosamines, some of which are potent carcinogens.

GREENHOUSE EFFECT

It is a heat-trapping process caused by gases such as carbon dioxide and water vapour which are transparent to incoming solar radiations but re-emit the infra-red radiations from the earth's surface. This helps to maintain the mean temperature at 15°C and were it not for this effect, the mean temperature would be at an unfavourable –18°C. What is causing a problem today is the enormous rate of increase in the concentration of CO_2, methane and chlorofluorocarbons (CFC) leading to the possibility of a climatic change of geological proportions occurring over time spans as short as a single human lifetime. This whole process has been dubbed 'the greenhouse effect" because the glass roof of the greenhouse lets sunlight in while blocking the passage of infrared radiations out.

PHOTOCHEMICAL SMOG

Stack gases emitted from industrial processes and automobile exhausts are toxic to plants. The term 'smog' is used for mixture of smoke and moisture in the air. The smog may be reducing or oxidising¾conversion of SO_2 to its acid aerosol is reducing while automobile exhaust gases, by the action of sunlight, form phytotoxic oxidants which are toxic to plants and this is oxidising 'smog'. Ozone and peroxyacetyl nitrate (PAN) are the two most important photochemical phyto-oxidants. Nitrogen dioxide and hydrocarbons of the automobile exhausts are energised by the ultraviolet light to react with oxygen resulting in production of ozone at a rate faster than its rate of decomposition to oxygen. Similar photochemical reactions between oxides of nitrogen and hydrocarbons result in the formation of peroxyacetyl nitrate (PAN). Photochemical smog is known to cause serious health hazards causing asthma and bronchitis which occur immediately after the smog.

CHLOROFLUORO CARBONS

'Miraculous refrigerants' used to replace the toxic and inflammable refrigerants like methyl chloride, ethyl chloride, propane and sulphur dioxide are marketed under the popular brand names such as 'freon' and 'genetron'. Fluorocarbons first came into use as refrigerants around 1930s because of their low toxicity, non-inflammability and least chemical reactivity. Later they became the standard propellants for dispersing aerosols.

Some chief chlorofluorocarbons are:

a. difluoro dichloro methane

b. fluorochloro methane

c. trichlorofluoromethane

d. dichlorofluoromethane

One of the important effects of chlorofluorocarbons is the depletion of the ozone layer. They do ten times greater damage to the ozone. The CFCs are very stable on the earth. But when they slowly drift up to the stratosphere, intense UV radiation acts on their chemical bonds releasing chlorine which strips an atom from the molecule of ozone turning it into ordinary oxygen. The chlorine goes on to repeat the process and in this way one CFC molecule can destroy thousands of molecules of ozone. There are about 200 processes by which CFCs can destroy the ozone. CFC lasts for an average of 74 years in the atmosphere and its concentration is doubling every seventeen years.

Review Questions

1. What is air pollution?
2. Describe in detail about the various gascous air pollutants.
3. Discuss particulate matter as an air pollutant.
4. What is greenhouse effect?
5. What is photochemical smog?

PART IV

MICROBES AND LITHOSPHERE

Relationship between Microbes and Land

Soil Formation

Soil Characterisation (Physical)

Soil Characterisation (Chemical)

Soil Microbiology

Soil Types and their Microflora

Quantification of Soil Microflora

Methods of Studying Ecology of Soil
Microflora

Role of Microbes in Soil Fertility

15

Relationship between Microbes and Land

The earth's crust consists of a wide variety of minerals which have been produced during several geological periods. Nearly 2000 minerals have been detected in the earth's crust but only a few constitute soil-forming rocks. These rocks were disintegrated to form the minerals which were decomposed to form the soil. The nature of the soil basically depends upon the nature of minerals present in them. Rocks are identified on the basis of minerals they contain. So the minerals should be studied before studying the soil forming rocks.

A mineral is a naturally occurring inorganic material that has fairly definite internal structure, composition and properties. Soil forming minerals have been broadly grouped as primary and secondary minerals. Primary minerals originate in the parent rock. Some of them remain unchanged during the course of soil formation. They are usually present in the sand fraction. Secondary minerals have formed by the alteration and decomposition of primary minerals. They are also called clay minerals because they are the chief constituents of clay.

In the earlier stages of soil development, soils were dominated by characteristics which they inherited from the parent material. They are dominated by the acquired characters at the later stages of soil development. For example, soils which were developed from the basic parent material were rich in basic elements and alkaline in reaction during the earlier stages of soil development. Later on these basic elements were gradually washed down by high rainfall and the soils ultimately became acidic at the later stages of their development.

The fertility of soil depends not only on its chemical composition, but also on the qualitative and quantitative nature of microorganisms inhabiting it. The microbes inhabiting soil can be classified into bacteria, actinomycetes, fungi, algae and protozoa and the branch of science dealing with these microbes and their activities in soil is called soil microbiology. Unlike soil science whose origin can be traced back to Roman and Aryan times, soil microbiology emerged as a distinct branch of soil science only in 1838.

Soil is inhabited by diverse groups of living organisms which include both macro- and micro- fauna and flora. The total number of living organisms in soil may be as much as billions per gm of soil and their live weight may go up to five tons per acre furrow slice of soil. Soil organisms depend mainly upon the climate and the resultant vegetation. Soil flora and fauna also depend upon soil factors like temperature, moisture – air relationships, soil reaction and humus and nutrient content of soils. Microflora are responsible for about 80% of the total soil metabolism. It decomposes organic residues to form the humus.

Lithosphere habitats occur as land masses consisting of rocks and soil. Soil is capable of acting as a habitat for biological organisms. Soil is a very complex environment composed of three main phases: solid, liquid and gases. Soil is formed by the physical, chemical and biological weathering of rocks to small particles.

Rocks are of three types:

- Igneous rocks formed by the solidification of molten lava.
- Sedimentary rocks formed by the deposition and consolidation of weathered products of other rocks.
- Metamorphic rocks formed by changes in form of other rocks.

Review Questions

1. What is a mineral?
2. What are clay minerals? Give examples.

16

Soil Formation

Soil is the outer covering of the earth which consists of loosely arranged layers of materials composed of inorganic and organic constituents in different stages of organization. It is the natural medium in which plants live, multiply and die and thus provides a perennial source of organic matter which could be recycled for plant nutrition. It provides the physical support needed for the anchorage of the root system and also serves as the reservoir of air, water and nutrients which are so essential for the biotic flora of the soil.

The process involved in the formation of soil are slow, gradual and continuous and are the sum total of environmental effects on rocks collectively known as the weathering of rocks.

Five important factors are involved in soil formation: (i) parent material, (ii) climate, (iii) topography, (iv) biological activity and (v) time.

Growth in soil is very slow or nil for much of the time. Most of the microbial population must be dormant in the agricultural soils perhaps except in the rhizosphere and at sites of local litter input. Mean generation times for soil bacteria may be in the region of 5-10 days. Rock surfaces provide a suitable habitat for a limited number of organisms. Some bacteria, algae, fungi and more specifically lichens colonise the terrestrial rock surfaces.

Bacteria and fungi are found in crevices which can retain water. Along the shores of oceans large population of cyanobacteria and algae inhabit rocky coasts. Some examples include *Calothrix*, Chlorophycophyta members like *Enteromorpha*, *Porphyra* (Rhodophycophyta).

Plant cover of the soil is an important factor in determining the types and number of microorganisms in that soil, since the plant root exudates and senescent parts of the plants are important sources of nutrients for soil microorganisms.

The extracellular polysaccharides that many microorganisms produce bind the individual soil particles together into crumbs. Hyphae of fungi and actinomycetes also play a part in this process. The crumb structure is very important in soil drainage and therefore in aeration. A soil with a good, well-developed crumb structure is freely draining and well aerated.

Microbes generally live on particle surfaces or in the interconnecting spaces (pores) between the crumbs. The distribution of soil microorganisms vary according to various factors like soil temperature, soil water, soil air, soil pH and soil organic and inorganic nutrients.

Bacteria in soil usually tend to grow as individuals or small microcolonies on the surface of the soil particles and roots.

In contrast, fungi grow in the soil from a food base. Hyphae grow, come upon some utilisable organic matter, cover the surface by branching and as the substrate is exhausted, hyphae grow off again.

SOIL MICROBIAL COMMUNITIES

In general soil microbial communities are divided into three broad groups:

Autochthonous   These are part of the soil microbial community capable of utilising humic substances (insoluble materials). They exhibit slow state of activity but are continuously growing (ascomycetes, basidiomycetous fungi, actinomycetes and some gram negative rods).

Zymogenous/opportunistic   They are organisms which can utilise only simpler often soluble substrate. They exhibit higher levels of activity and rapid growth on easily utilisable substrates that are available in the form of fresh plant litter, animal droppings and fresh carcasses. Intermittent activity and inactive resting stages are characteristic of such zymogenous organisms, e.g. *Pseudomonas, Bacillus, Penicillium, Aspergillus* and *Mucor*.

Allochthonous   Another community which is quite different from the zymogenous is the population which is totally foreign to the soil. For example pathogens of humans, animals and plants that do not find suitable conditions in the soil to survive get deposited in the soil from various sources and are temporary dwellers of soil. They perish soon if a suitable host is not within their reach. In other words, those organisms which cannot survive in their vegetative stage in the soil (since soil is not their true environment) and those which do not have any adaptations (like spores) to tide over the unfavourable environment are termed as allochthonous (foreign).

SOIL MINERALS

The nature of soil basically depends on the nature of minerals present in it. Minerals are naturally occurring inorganic material that have definite structure, composition and properties. Soil minerals are of two types:

Primary minerals which originate in the parent rock, e.g. feldspars, pyroxenes, amphiboles, mica, etc.

Secondary minerals which are formed from the alteration and decomposition of primary minerals (also called *clay minerals* since the chief constituent in them is clay), e.g. silicate clays like kaolinite, smectite, vermiculite.

Soil formation is a function of climate, topography, parent material, time and biota. Two particular soils may have the same parent material but develop unique characteristics depending on their environment.

WEATHERING PROCESS

Changes, disintegrations and decompositions caused by the action of the atmospheric agents on rocks and minerals at or near the surface of the earth is called the weathering process.

Weathering is of three types:

- physical or mechanical weathering
- chemical weathering
- biological weathering

Mechanical Process of Weathering

Rocks heat up during the day and cool down during the night. Different minerals present in the rocks expand (day) and contract (night) at different rates (since the minerals differ in their coefficient of expansion, some expand and contract more than others). This differential expansion and contraction of minerals is a continuous process leading to cracks on the rock surface.

Expansion of Rocks

During the day, heat is conducted very slowly from the outer to the inner layer of the rock. Hence outer layer is at a higher temperature (expands more than the inner layer during the day) than the inner layers. The same experiences greater temperature drop during the night (i.e. surface layers are cooler than inner layers at night). This process of warming and cooling continues till the outer layer gets separated from the inner layer (exfoliation). During winter, water freezes in the cracks to form ice leading to tremendous pressure exertion on the rock (due to expansion of ice). Cracks widen due to this pressure. When ice melts (warm season) and again freezes (cold season), it leads to widening of cracks and ultimately causes the rocks to gradually break down to smaller and smaller pieces. Rocks can also be disintegrated by the abrasive action of glaciers¾water running down the rocks carry away small pieces of rocks with them causing cutting of these rocks (water loaded with rocks have tremendous cutting power). Wind action also takes part in mechanical weathering by scraping rocks and carrying away fine particles.

Chemical Process of Weathering

Atmospheric gases like CO_2, SO_2, N_2, O_2 and water vapour react with rocks and minerals and change them to new minerals. The following reactions are important as chemical weathering agents.

Hydrolysis When minerals react with water:

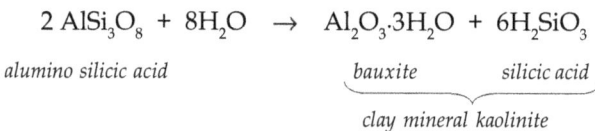

$$2\ AlSi_3O_8 + 8H_2O \rightarrow Al_2O_3 \cdot 3H_2O + 6H_2SiO_3$$

alumino silicic acid   *bauxite*   *silicic acid*

clay mineral kaolinite

Hydration When H^- and OH ions get attached to minerals to become an integral part of them.

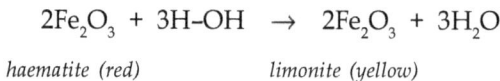

$$2Fe_2O_3 + 3H\text{--}OH \rightarrow 2Fe_2O_3 + 3H_2O$$

haematite (red)   *limonite (yellow)*

Carbonation When carbonic acid reacts with minerals and changes them to new minerals.

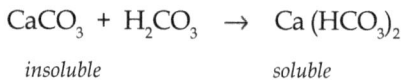

$$CaCO_3 + H_2CO_3 \rightarrow Ca\,(HCO_3)_2$$

insoluble   *soluble*

Solution Rain water gradually dissolves the insoluble minerals. For example, heavy rain washes the Ca and carbonate ions down thus solubilising calcium carbonate in water.

Oxidation Process by which electrons are lost from ions (mineral ions) whose positive charge is increased.

For example,   Fe''' ion $\rightarrow$ Fe'' ion

Even when a compound combines with oxygen, it is said to be oxidised. For example, when Fe'' ion in ferromagnesian mineral get easily oxidised to Fe''' ion. During this process, some positive ions go out of the crystal lattice of mineral in order to maintain electrical neutrality (since electron have been removed) leaving some portion of the crystal lattice of the mineral empty, facilitating entry of ions into it and exit of some ions from it. Hence ferromagnesian minerals are rapidly decomposed this way.

Biological Weathering

This involves disintegration of bare rocks by the action of living organisms, thus adding to the formation on the soil.

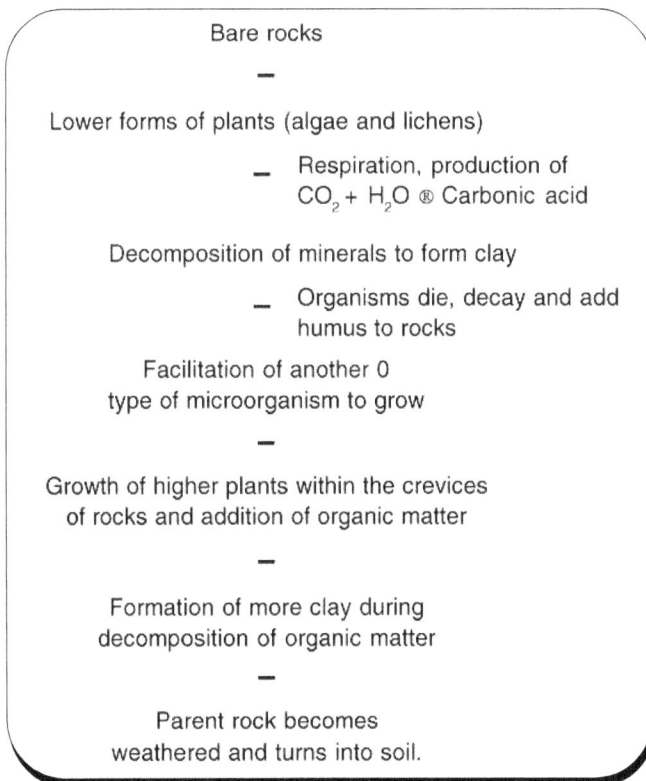

Bare rocks

—

Lower forms of plants (algae and lichens)

— Respiration, production of
CO_2 + H_2O ® Carbonic acid

Decomposition of minerals to form clay

— Organisms die, decay and add
humus to rocks

Facilitation of another 0
type of microorganism to grow

—

Growth of higher plants within the crevices
of rocks and addition of organic matter

—

Formation of more clay during
decomposition of organic matter

—

Parent rock becomes
weathered and turns into soil.

Lower forms of plants (algae and pioneer organisms (lichens inhabit the bare rocks, since lichens can grow in any extreme habitat owing to their symbiotic relationship.

These organisms release carbon dioxide (respiration) which combines with water (moisture) to form carbonic acid.

The acid helps in decomposing the soil minerals to form clay. This is a long-term process. As these pioneer organisms die, they get decomposed and add humus to the rock surface thus facilitating other chemoorganotrophs to grow. Now there is a heap of life growing over the rock and this leads to the formation of crevices on the rocks (due to the acid produced by the growth of the organisms). Higher plants start growing in these crevices adding to the organic matter. Thus the parent rock becomes weathered and turns into soil. This process happens only in terms of years.

Soil Genesis

Higher plants begin to grow and add organic matter to the soil which is further decomposed to form humus that combines with clay to form clay-humus complex (surface of the soil is thus dark). This process continues till a dark layer of soil, about a foot in thickness is developed. Thus a horizon is formed (Fig. 16.1).

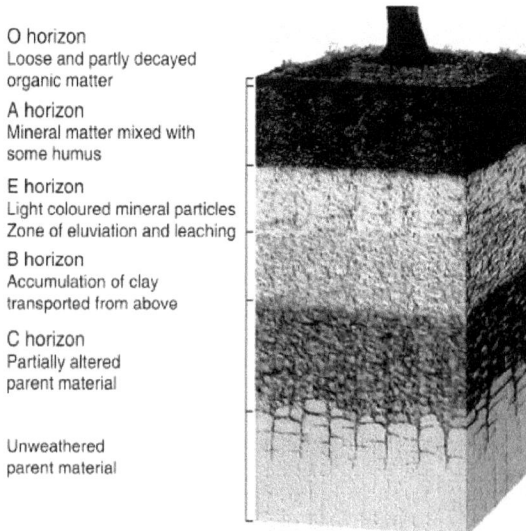

O horizon
Loose and partly decayed
organic matter

A horizon
Mineral matter mixed with
some humus

E horizon
Light coloured mineral particles
Zone of eluviation and leaching

B horizon
Accumulation of clay
transported from above

C horizon
Partially altered
parent material

Unweathered
parent material

Figure 16.1   *Soil profile*

The various steps involved in the soil formation during physical, chemical and biological weathering processes are shown below:

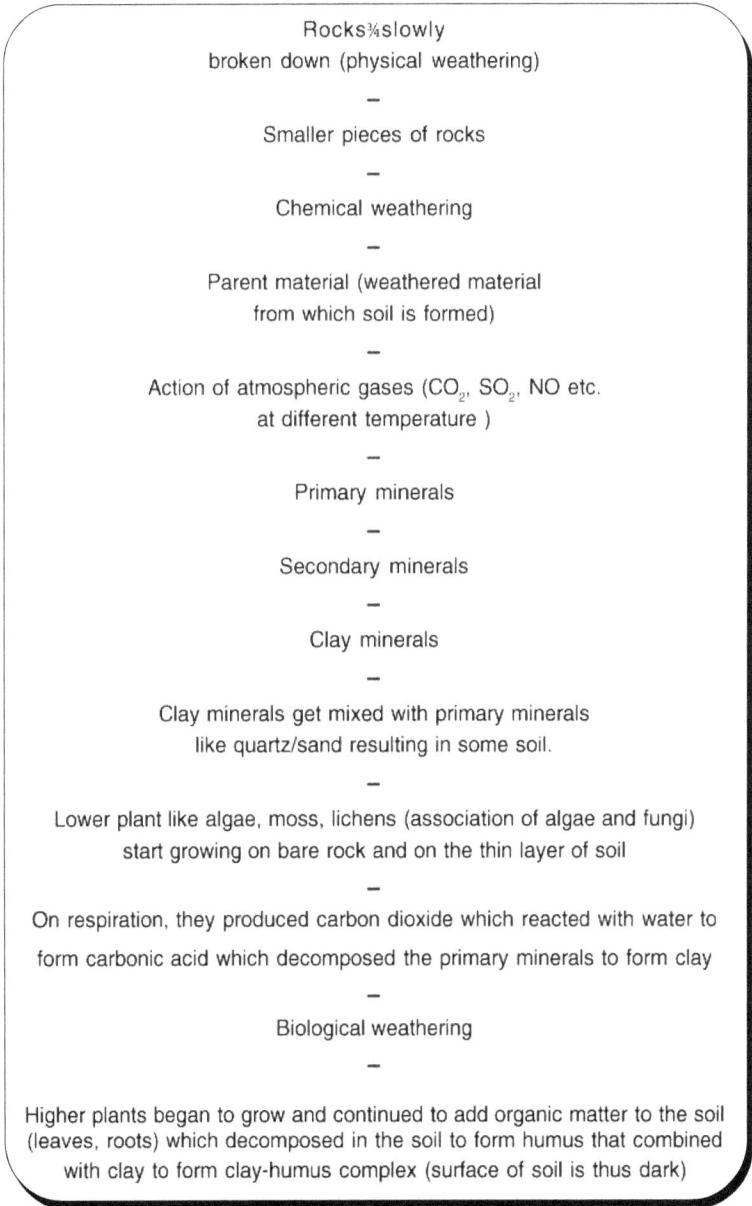

Rocks¾slowly
broken down (physical weathering)

–

Smaller pieces of rocks

–

Chemical weathering

–

Parent material (weathered material
from which soil is formed)

–

Action of atmospheric gases (CO_2, SO_2, NO etc.
at different temperature)

–

Primary minerals

–

Secondary minerals

–

Clay minerals

–

Clay minerals get mixed with primary minerals
like quartz/sand resulting in some soil.

–

Lower plant like algae, moss, lichens (association of algae and fungi)
start growing on bare rock and on the thin layer of soil

–

On respiration, they produced carbon dioxide which reacted with water to
form carbonic acid which decomposed the primary minerals to form clay

–

Biological weathering

–

Higher plants began to grow and continued to add organic matter to the soil
(leaves, roots) which decomposed in the soil to form humus that combined
with clay to form clay-humus complex (surface of soil is thus dark)

The parent material is at a considerable distance from the surface hence cannot be subjected to the direct action of atmospheric agents (heat, moisture, gases, etc). Hence it continued to decompose at a slower rate forming a compact lighter layer called the B horizon.

Some clay was gradually washed down (eluviated from) the lower portion of A horizon (the E horizon) and was washed in (illuviated) the B horizon.

During the early stage of soil development, soils are dominated by characteristics inherited from parent material. Later on they show a change in their chemical characteristics. For example, an originally basic soil (obtained from basic primary mineral) gets washed off its basic elements and ultimately becomes acidic in reaction.

Not every soil has all the horizons. Young soils may simply have A and C horizon. Mature soils have more horizons.

Factors affecting the weathering process include (a)climate, (b) physical characteristics and (c) chemical composition of the rocks.

Review Questions

1. Discuss the soil microbial communities.

2. Give a detailed account of the weathering process.

3. Discuss the role played by biotic community in soil formation.

4. Give an illustrated account of soil profile.

17

Soil Characterisation (Physical)

Physical properties of soil type depend on the size of particles in it. Soil particles occupy roughly more than half the space in the soil. The remaining space between the particles, called the pore space, is occupied by water and air. The pore size or porosity of soils together with bulk density determine the structure of soils. The stability of soil aggregates depends on the organic matter content of individual soils and the nature of microbial products which binds particles together.

Soil is made up of five major components: mineral matter, water, air , organic matter and living organisms. Soil characterisation can be achieved by taking a look at the physical, chemical and biological properties of soil.

PHYSICAL CHARACTERISATION OF SOIL

Those properties which can be evaluated by observing and feeling the soil are called physical properties and can be measured. For example, size, density, temperature, water, etc. are physical properties. Here, no chemical changes are involved. Let us discuss the properties one by one.

Mechanical Composition of Soil

There are different kinds of soil based on the composition of the inorganic particles that make them up. The International system classifies the mineral composition make-up as follows:

Soil particle/separates	Diameter in mm
Coarse sand	2 – 0.2 mm
Fine sand	0.2 – 0.02 mm
Silt	0.02 – 0.002 mm
Clay	< 0.002 mm

Nature of soil separates   Quartz dominates the grades of sand. Apart from quartz, mica, feldspar/calcite may also be present.

Soil Texture

Texture refers to size of the soil particles. Texture is the relative proportion of sand, silt and clay for each soil. Texture classes are:

- *Coarse texture*   Sands, loamy sand and sandy loams with less than 18% clay, and more than 65% sand.

- *Medium texture*   Sandy loams, loams, sandy clay loams, silt loams with less than 35% clay and less than 65 % sand; the sand fractions may be as high as 82% if a minimum of 18 % clay is present

- *Fine texture*   Clays, silty clays, sandy clays, clay loams and silty clay loams with more than 35% clay.

Texture is a basic physical property of soil and is one of the major features for differentiating kinds of soils. The extremes of soil texture are not particularly desirable for most uses. Generally, soils of medium texture (loams) would be most desirable, especially for plant growth. Texture is an inherited property of soil and is not changed by agricultural cultivation except when the topsoil might be mixed in with the subsoil during deep ploughing or when some of the topsoil has been eroded away.

Importance of soil texture   Capacity of soils to store water and nutrients increases when their clay percentage increases since clay has a large surface area/unit volume and they can absorb large amounts of nutrients and water. A soil with sandy texture has difficulty in retaining water and thus nutrients are not made available to the plants growing in such a soil. A clayey soil is poorer in aeration (due to stagnation of water) thus debilitating the plant growth. Loamy soil (a mixture of clay and sand) is the best textured soil for crop cultivation since it has all the beneficial aspects not found in the sandy and clayey soil. Thus we can see how texture affects the plant growth.

Soil Pore Space

Pore space is that portion of the soil volume which is not occupied by soil solid but by air and/or water. Pore space in soil is of two kinds: the macro pore space which has a diameter of more than 60 mm stores exchangeable air and the micro pore space which has a diameter of less than 30 mm stores capillary water. The texture, organic matter present in soil, nature of crops cultivated and the soil depth have a great influence over the soil pore space.

Soil Colour

Each soil has a definite colour of its own which is mainly due to the constituents present in it. If the soil is rich in organic matter (humus which is dark in colour) then the soil is darker in nature and vice versa. If the soil is rich in some inorganic matter like ferric oxide (red) then it takes up the colour of the inorganic matter. Soil colour tells about the soil condition and the necessary remedial measures ought to be taken to improve the soil. For instance, black soil indicates fertility and light coloured soil indicates lack of fertility. Soil colour is influenced by the organic matter present in it and the depth of soil.

Soil Aeration

Major gases in the atmosphere are also found in soil. There are three important gases in the soil. They are nitrogen (79%), oxygen (18–20%) and carbon dioxide (1–10%). Soil CO_2 is higher than atmospheric CO_2. It is a mechanism of rapid exchange of oxygen and carbon dioxide between soil pore space (macro space) and the atmosphere in order to prevent the deficiency of oxygen and toxicity of carbon dioxide in the soil air. Compact soils of fine texture may suffer from poor aeration due to water logging (gaseous exchange may not be so rapid to remove CO_2 from soil air and to supply oxygen to the roots). This also happens when there is an excessive amount of readily decomposable organic matter added to the soil. Ploughing the soil during cultivation may prevent such a condition.

Soil aeration is accomplished by two methods: mass flow and diffusion. During mass flow, there is a movement of air *en masse* in between the atmosphere and the soil, i.e. during the day, soil is hotter than the atmosphere and the soil gases expand and pass to the atmosphere rapidly. During the night the soil gets cooler than the atmosphere and the gases flow into the soil *en masse* from the atmosphere. Most of the gaseous exchange in the soil takes place by diffusion mechanism. Each gas in the atmosphere exerts its own partial pressure in relation to the volume of air. Atmosphere has a high amount of exchangeable oxygen than soil, thus the partial pressure of oxygen is more in the atmosphere than in the soil. Hence there is a movement of oxygen from the atmosphere to the soil (gases move from an area of high partial pressure to an area of low pressure). In contrast, the level of carbon dioxide in soil is more (due to the respiratory release of the soil micro-and macro flora and fauna) than that found in the atmosphere. Hence there is movement of this gas from the soil to the atmosphere. Thus there is a displacement of both the gases between the soil and the atmosphere thus facilitating gaseous exchange. Organic matter, depth and soil moisture have a role to play in influencing the soil aeration.

Soil Temperature

Every biological process needs an optimum temperature to get accomplished. A maize seed germinates only at a temperature range of 7-10°C and most of the soil microbes function best at 25-35°C Soil gets heated up mainly due to solar radiation and its temperature is highly influenced by the amount of solar radiation received by the soil which in turn depends on the climate of the region. The amount of sunlight reaching the soil again depends on the slope of the land. For example, if the land is slopy, then the amount of solar radiation striking the unit volume of land decreases as the slope increases. This is not the case in the leveled land where each unit area of land receives equal proportion of solar radiation. Soil temperature also depends on the vegetation cover of the land. A barren land gets heated up faster and cools up at a rapid rate whereas in a land covered with vegetation which acts as an insulation barrier, soil temperature remains near optimum. Soil colour also influences the temperature as seen in darker soils that absorb more solar radiation than a light coloured soil.

Soil Water

Water is a very important constituent of soil since all biological phenomena need water. Nutrients are dissolved in water and can be made available to the plant roots only in the form of soil solution. Soil water regulates physical, chemical and biological properties of soil. Major portion of water absorption by the plant roots is through the soil.

The sources of soil water include rainfall, rivers, wells and ponds. The different forms of soil water include the following:

Gravitational water It is that water which moves under the influence of gravity down the macropores. It is harmful to the plants since it removes the nutrients from the soil.

Capillary water It is that water that binds at the soil surface and is directly available to the plant roots since it is found within the micropore space in the soil.

Hygroscopic water It is in the form of vapour and is not available for plant growth.

SOIL STRUCTURE

Soil structure is not a single entity. It is a complex of various physical characteristics of soil, like size of soil particles, pore size, hydraulic conductivity of soil, bulk density, etc.

> **Important Terminology**
>
> ***Soil aggregates*** Primary soil particles, sand, silt and clay are united together to form secondary soil particles which are called soil aggregates.
>
> ***Ped*** A naturally occurring soil aggregate.
>
> ***Clod*** Formed artificially by ploughing a dry clayey land soil.
>
> The arrangement of peds with respect to each other determines soil structure.

Types of Soil Structure

Soil structure is the shape that the soil takes based on its physical and chemical properties. Each individual unit of soil structure is called a ped. Take a sample of undisturbed soil in your hand (either from the pit or from the shovel or auger). Look closely at the soil in your hand and examine its structure. Possible choices of soil structure are:

Granular Resembles biscuit crumbs and is usually less than 0.5 cm in diameter. Commonly found in surface horizons where roots have been growing.

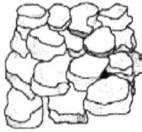

Blocky   Irregular blocks that are usually 1.5–5.0 cm in diameter.

Prismatic   Vertical columns of soil that might be many centimetres long. Usually found in lower horizons.

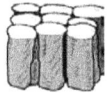

Columnar   Vertical columns of soil that have a salt cap at the top. Found in soils of arid climates.

Platy   Thin, flat plates of soil that lie horizontally. Usually found in compacted soil.

Single grained   Soil is broken into individual particles that do not stick together. Always accompanies a loose consistency. Commonly found in sandy soils.

Massive   Soil has no visible structure, is hard to break apart and appears in very large clods.

Grades of Soil Structure

This refers to the durability of peds.

Structureless   No peds are observed in the undisturbed soil.

Weak   Poorly formed indistinct peds are found in undisturbed soil.

Moderate   Fairly well-formed peds.

Strong   Well-formed distinct peds.

Formation of Soil Aggregate

Water molecules are dipolar. One side of the water molecule is positively charged while the opposite side is negatively charged. The negative end of water molecule attracts the positive end of the second water molecule thus forming long chains of water molecule. Cations get attracted to the negative end of water chain, the positive end of which are attracted by clay particle thus holding the clay particles by bridges of water molecules. When soil dries up, the chain of water molecule shortens more and more; ultimately clay particles are united by some organic compounds formed during the decomposition of soil organic matter. Soil microorganisms decompose organic matter to form dark gummy substance which bind the clay, sand and silt particles to form soil aggregate. Fungal hyphae and root hairs also bind the particles of soil to form aggregates.

Role of Organic Matter in Soil Structure

Soils receiving well-decomposed organic manures have better soil aggregates than those receiving not so easily decomposable organic matter. The exudates of roots along with sloughed off debris of root cortex help in creating soil aggregates. Microbial products such as gums are the deciding factors in the improvement of soil structure. The complex polysaccharides of microbial origin are resistant to microbial attack. Among the carbohydrates isolated from soil, dextran containing high amounts of uronic acid which is resistant to microbial degradation has the best soil aggregating qualities.

Improvement of Soil Structure

Addition of lime to acidic soils improves their structure because lime stimulates growth of microbes (responsible for the improvement of soil structure). Addition of calcium nitrate also improves soil structure. Addition of large doses of partially decomposed organic manures like FYM, compost and green manure crops greatly improves soil structure. Surface of land covered with waste organic material like grasses, leaves, straw, etc. (mulches) keep the soil surface moist and cool so that microbes can multiply rapidly and improve soil structure. Mulches also protect the soil surface from the beating action of the raindrop that may destroy the soil structure to a great extent. There are some synthetic soil conditioners like polyacrylic acid (PAA), polyacrylonitrile (PAN) and vinyl acetate maleic acid copolymer (VAMA). But they are very costly.

Evaluation of Soil Structure

Soil structure is determined by the following factors:

Size of the soil particles   Large-sized particles like sand or coarse sand do not form stable soil aggregates. Particles of size less than 0.002 mm like clay play a major role in building up the soil structure.

Pore size   The micropore size affects the soil aggregate formation. As the pore size decreases, water retention capacity increases. Thus in clayey soil, there is a greater chance of soil aggregate formation than in a sandy soil.

Saturated flow of water   Soil is saturated with water, i.e. even the macropores are full of water. Water moves in the soil according to the difference in potential (water flows from a higher potential to the state of lower potential, i.e. drier soil). This potential difference is called the hydraulic conductivity of soil and depends on macropores and viscosity of water.

Bulk density of soil   It is the weight of dry solids per unit volume of soil which is the sum of volume occupied by soil solids and air.

$$BD = \text{Wt. of soil solids}/(\text{Vol. of soil solids} + \text{Vol. of air})$$

Dry soils of finer texture (less soil particle diameter) like clay have more space filled with air than those of coarse texture. Hence weight per unit volume of soil is reduced, reducing the bulk density of soil.

Importance of Soil Structure

A well-structured soil increases water intake and drainage and has a direct effect on plant growth. It influences biological activities such as nitrogen fixation, nitrification and decomposition.

Review Questions

1. Define soil texture and highlight its significance.
2. Discuss in brief about soil aeration and its mechanism.
3. Give an account of soil water.
4. Define
   a. ped
   b. clod
   c. aggregate
5. Describe the various structures of soil.
6. Discuss the formation of soil aggregate.
7. Brief out the role of organic matter in maintaining soil structure.
8. How can you improve soil structure?
9. Describe in detail the various physical parametres involved in maintenance of soil as a habitat.

18

Soil Characterisation (Chemical)

S oil is the medium from which plants normally derive their nutrients. The three main components of soil which provide nutrients for plant growth are the organic matter, the derivatives of parent rock materials and the clay fraction. The organic matter in soils is a potential source of N, P and S for plant growth.

CHEMICAL PROPERTIES OF SOIL

The chemical properties of soils refers to the nature of the chemical changes taking place in them which in turn depend upon their chemical compositions and the nature of the inorganic and organic materials contained in them, which have originated from the gradual decomposition of the soil and organic materials, mainly of plant origin.

Chemical Composition of Soil

Soil is mainly made up of oxygen (46.7%), silicon (27%), aluminium (8.1%) and iron (5.0%). Plant nutrients like Ca, Mg, K, Na, P and S are present in the minerals and in the soil solution. O_2, Si, and Al occur as constituents of minerals and as oxides. Fe occurs mainly in the form of oxides and ferromagnesium minerals. Ca occurs mainly in calcite, gypsum, apatite and dolomite. Mg is present mainly in dolomite and hornblend. K occurs mainly in microcline and mica. P occurs as aluminium phosphate and calcium phosphate and in the organic form as phospholipids, inositol, choline, etc. N occurs mainly in the organic form as proteins, amino acids, etc. All micronutrients like Mo, Fe, Mn, Zn, Cu, B occur in the inorganic form.

Soil Colloids

Most of the inorganic and organic matter in soil is in the form of minute particles, with a diameter of about 0.00002 mm and can remain suspended in water for a long period. They have an enormous surface area and are negatively charged, hence they repel each other in the suspension. These inorganic and organic particles in soil are called soil colloids (*colla* meaning glue-like).

Soil colloids are of two kinds:

- **Inorganic colloids** which exist mostly as silicate clay found in less weathered soils in the temperate regions and as hydrous oxides of iron and aluminium clay in the extremely weathered soils in the tropics.
- **Organic colloids** which are represented by humus.

Soil Humus

Soil humus is a mixture of dark, colloidal organic compounds relatively resistant to decomposition. These compounds result from the decay of organic litter and accumulate in the O and A horizons of soils. Humus includes sugar amines, nucleic acids, phospholipids, vitamins, sulpholipids, polysaccharides and many other unclassified compounds.

Formation of Humus

Humus is formed during the decomposition of organic "litter" (including pine needles, leaves, and animal droppings) in soils. This decay is mediated by microbes and the enzymes they excrete (which break certain specific bonds in organic matter). The main reactions of this decomposition are:

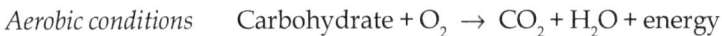

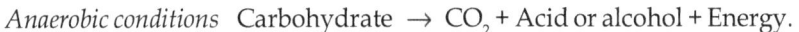

Aerobic conditions   Carbohydrate + O_2 $\rightarrow$ CO_2 + H_2O + energy

Anaerobic conditions Carbohydrate $\rightarrow$ CO_2 + Acid or alcohol + Energy.

These reactions are much complicated by the complex structure of the litter, and the products of decomposition include various nutrients, organic acids and amines (depending on the conditions and starting materials), in addition to the "resistant residues" or humus. Between 60 and 80 per cent of the carbon from most plant residues is evolved as CO_2 within a year of deposition; 5 to 15 per cent is incorporated into the microbial biomass and the rest remains in soil humus. Water-soluble, readily available substrates such as sugars, amino acids and pyrimidines are quickly metabolised during decomposition (generally within a few hours or days) while less reactive compounds are digested more slowly. This imbalance leads to accumulation of more recalcitrant compounds (i.e. humus) in the soil.

Early in decomposition, simple, phenolic compounds, bits of lignin and melanin (lignin-like molecules) and other phenolic polymers are transformed by b-oxidation of side chains, addition of hydroxyl groups, oxidation of methyl groups and decarboxylation. The more reactive compounds are oxidised and then converted to radicals, which stabilise by linking into dimers or forming quinones (oxidised polyphenols). The less reactive compounds add onto the polymers through nucleophilic addition to quinones. These linkages repeat to create humic macromolecules. During the decomposition or humification of organic litter, the carbon to nitrogen (C/N) and carbon to sulphur (C/S) ratios decrease, indicating that relatively more carbon than nitrogen or sulphur is "lost" in the process. Numerous factors control or influence the decomposition of organic matter. Among these are the properties, amount and stage of decay of the organic matter and the availability of oxygen, temperature, soil moisture, nutrients and soil texture. As acidity increases, soil respiration decreases, leading to an accumulation of organic matter and a drop in the rate of litter decomposition. When soil moisture was considered, the presence of live roots stimulated the decomposition of organic material because roots increased microbial activity.

In summary, through a series of complex reactions, microbes mediate the conversion of organic material such as leaves and twigs to the dark humus which colours and glues soils. Without these processes, the soils would be loose, non-cemented dusts and no life would be able to grow on them, and the world would be a very different place.

Cation Exchange in Soils

The soil humus consists of a negatively charged anion called a *micelle*. Each negative charge on the humic micelle attracts a univalent cation. The humic micelle is made up of carbon, hydrogen and oxygen. Sources of negative charges on the humic micelle are a partially dissociated carboxylic group (-COOH), enolic hydroxyl group (-OH) and phenolic hydroxyl groups (-OH). The negative charge on a humic micelle depends on the pH. When the pH increases, more H+ ions dissociate from the groups. Consequently the negative charge on the humic micelle increases. Clay matter of the soil is a mixture of sand and humus. There are cations surrounding the clay micelle which are called exchangeable cations because they can be replaced by other cations. When ammonium sulphate is added to the soil, the ammonium ions gradually replace other cations especially calcium ions. This process is reversible and is called cation exchange.

Cation exchange capacity (CEC) is the total number of negative charges per unit weight of the soil. In other words, it is the total number of cation adsorption sites per unit weight of soils.

The factors affecting cation exchange capacity include the following:

Soil texture   The CEC of soils increases when their percentage of clay increases i.e. when their texture becomes finer.

Soil humus content   Since a negatively charged humic micelle attracts cations and holds them, the CEC increases when the percentage of humus increases.

Nature of clay   The CEC depends on the nature of clay minerals present in it since each mineral has its own capacity to exchange and hold cations. The CEC of a soil dominated by vermiculite is much higher than the CEC of another soil dominated by kaolinite.

Soil reaction   When the pH of soil increases, more H+ ions dissociate from the clay minerals especially kaolinite, thereby the CEC of soil dominated by kaolinite also increases.

Review Questions

1. Discuss the chemical properties of soil.
2. What is humus? Describe its foundation.
3. What is cation exchange capacity? Discuss the factors affecting the cation exchange capacity?

19

Soil Microbiology

S oil is the loose outer material of earth's surface, a layer which is different from the underlying bedrock. Agriculturally, it supports plant life. Chemically, it contains a multitude of organic and inorganic substances not found in the underlying strata. Microbiologically, it contains a vast array of microorganisms representing a dynamic site of biological interaction. Thus soil is a consortium (mixture) of living cells in an organic or mineral matrix. Neither the living cells nor the composition of the organic/inorganic matrix are constant, both vary with time and location. Thus soil microbiology is the study of the microflora and microfauna of soil.

MAJOR GROUPS OF SOIL MICROORGANISMS

Soil contains five major groups of microorganisms. Bacteria, fungi, actinomycetes, algae, protozoa and viruses form a small portion of soil microflora. Basically, the soil organisms are classified as

Autotrophs (utilise the inorganic minerals)	Heterotrophs (utilise the organic matter)
Main producers	Consumers & Decomposers
photoautotrophs	photoheterotrophs
chemoautotrophs	chemoheterotrophs

The factors that influence microbial distribution may be intrinsic or extrinsic.

Intrinsic factors arise from the structure and function of the microbes themselves, through their persistence in the soil (as spores), their size, motility, structural features (stalk, holdfast, filaments) and biochemical capacities. Extrinsic factors are those arising from soil and the environment like pH, temperature, water, redox potential, radiation, etc.

QUALITATIVE MICROFLORA OF SOIL

One gram of soil contains various groups of microflora belonging to all sections of prokaryotic and eukaryotic groups.

Bacteria

They are the most abundant group, their numbers ranging from 100,000 to 100 million/g of soil. They account for almost all the biological and chemical changes in the environment containing little or no oxygen.

Based on ecological differentiation, soil bacteria can be classified as autochthonous and zymogenous bacteria.

Autochthonous bacteria are indigenous/native of soil and their population is uniform.

Zymogenous bacteria are invaders of soil and their population depends on external source of energy. Their normal population is low, but when a specific substrate is added to the soil, they increase in number.

Based on their capacity to utilise oxygen, they are classified as aerobes, e.g. *Bacillus* sp., *Pseudomonas* sp. and anaerobes, e.g. *Clostridium* sp.

Based on nutritional requirements, soil bacteria are classified as:

■ those requiring amino acids

■ those requiring B–vitamins

■ those requiring both

The most common bacterial flora of the soil are *Pseudomonas* sp., *Arthrobacter* sp., *Flavobacterium* sp., *Sarcina* sp., *Enterobacter* sp., *Bacillus* sp., etc.

Factors affecting bacterial population in soil

Moisture   It governs microbial activity in two ways¾adequate supply is necessary (since water is the major constituent of protoplasm) but excessive moisture prevents microbial proliferation by limiting gaseous exchange and lowering oxygen supply thereby decreasing aerobic bacterial counts.

Temperature   Each bacterium has an optimum temperature for growth. Most are mesophiles preferring the range of 25–35°C. Certain species develop best at temperatures less than 20°C. They are called psychrophiles. Thermophiles are ubiquitos at temperatures from 45°C–55°C.

Organic matter Bacterial size is directly related to organic matter. Humus rich localities are richest in bacteria. Addition of carbonaceous materials and ploughing down of green manure or crop residues have a positive effect on the bacterial counts in soil.

pH Highly acidic or alkaline conditions inhibit soil bacteria.

Inorganic nutrients This commonly increase the bacterial numbers but ammonium fertilisers tend to suppress bacterial flora due to the acidity generated through the microbial oxidation of ammonium to nitric acid.

Cultivation practices Ploughing and tillage operations increase the number of soil bacteria by improving soil structure and porosity favouring aeration and altering the moisture status and exposing inaccessible organic nutrients to bacterial action.

Seasons They play a minor role. In temperate regions, a burst of activity occurs in spring when soil becomes warm and rich in organic matter from the previous autumn. Thus the highest count of bacteria are seen in spring and autumn. Bacterial counts decline in hot summer and winter where cells remain in a state of biochemical inactivity.

Depth It is a secondary and variable factor. Bacteria are found more in the upper few centimetres. At the very surface, the bacterial community is less due to inadequate moisture and the bactericidal activity of the sun. The highest number is found at 1–2 cm from the surface and therefore the number declines with depth.

Functions All major mineral transformations or biogeochemical cycling of nitrogen, carbon, sulphur, phosphorus, iron, manganese and other trace elements are due to the activities of the soil microflora.

Mineralisation and immobilisation of minerals, production of growth hormones like gibberellins and auxins, and production of antibiotics and other commercially important products like enzymes are carried out by the soil microflora.

Actinomycetes

They are regarded as an intermediate group between bacteria and fungi. Like fungi, they produce slender branched filaments that develop into a mycelium. They also produce single, paired or a chain of asexual spores called conidia on their hypae. They do not show turbidity when grown in liquid cultures. Like bacteria some genera, show pressure of flagella, similarities in cell wall composition and sensitivity to antibacterial compounds and not to antifungals.

Actinomycetes grow slowly and are present in the surface soil and also in the lower horizons particularly in environments of high pH.

Some members are *Streptomyces, Actinoplane, Frankia, Actinomyces, Thermomonospora,* etc.

Factors influencing actinomycete population in soil

Organic matter Actinomycetes are abundant in land having a high organic matter status. When the soil is amended with animal manure and organic manure, their number in the soil increases.

pH Actinomycetes are tolerant to acidity and their number declines below pH 5.

Temperature Most actinomycetes are mesophilic and thermophilic. Their numbers are high during composting (since heat is evolved during the decomposition of organic manner). They can tolerate up to 65°C.

Moisture Since most of them are aerobic, actinomycetes cannot withstand waterlogging. Unlike bacteria, they (especially the filamentous forms) can withstand dry conditions. Even in the deserts actinomycetes dominate the other forms.

Depth Filmentous forms found in the A horizon as well as at considerable depth make up a larger segment of the subsurface community, i.e. they are found even in the C horizon, e.g. *Nocardia, Streptomyces, Actinoplanes.*

Functions They are ineffective competitors since they are fastidious organisms. They are less in number when nutrients are high because of greater competition from bacteria and fungi.

They are heterotrophic feeders. Cellulose is decomposed by many species apart from proteins, lipids, starch and chitin.

Many species have the capacity to synthesise toxic metabolites like antibiotics which are effective against bacteria, fungi and yeasts.

Many species of *Streptomyces* liberate extracellular enzymes which lyse bacteria.

Actinomycetes participate in the following processes:

- Decomposition of certain resistant components of plant and animal tissues. They are effective competitors only when resistant compounds remain.

- Formation of humus through the conversion of plant remains and leaf litters.
- Transformation at high temperature. For example, composting and manuring by thermophilic actinomycetes will be common.
- Actinomycetes cause certain plant diseases, e.g. potato scab.
- They also cause certain human infections like Nocardiosis.
- They are important as microbial antagonists and in regulating the composition of soil community.

Soil Fungi

In most well-aerated soils, fungi are present in large numbers. They are present in the form of mycelial bits, rhizomorphs or spores. All fungi are heterotrophic in nutrition and dominate acid soils. Some can tolerate pH beyond 9.0 also. In the lab, fungal cultures can be obtained in pure form by acidifying the media or adding bacteriostatic agents like rosebengal and streptomycin.

Some important fungal genera include *Aspergillus, Mucor, Rhizopus, Penicillium, Fusarium, Curvularia, Alternaria, Helmithosporium, Cladosporium* etc.

Factors influencing the fungal population in soil

Organic nutrients Fungal distribution is determined chiefly by the availability of oxidisable carbonaceous substrates. During decomposition fungal population is the greatest at the initial stages.

pH Fungi are common in acid habitats because of lack of competition by other microbes for food reserves.

Inorganic nutrients Nitrogen-containing fertilisers favour an increase in fungal population since their addition results in acidifying the soil.

Moisture Their capacity for catalysing chemical changes is poor at low water supply.

Temperature Most species of fungi are mesophilic. Thermophilic growth is not common, e.g. *Aspergillus, Mucor,* etc.

Cultivation practices Fungi are more common in cultivated lands than in uncultivated lands.

Depth   In cultivated soils, fungi are more common in surface layers but high counts are also found in the B horizon.

Functions and activity of fungi

- Fungi contain no chlorophyll, hence they must obtain carbon for metabolism from preformed organic molecules. Among carbon sources used up are sugars, organic acids, starch, pectin, cellulose, fats and lignin that are particularly resistant to bacterial degradation.

- Fungi act as predators devouring protozoans and nematodes, e.g., *Arthrobotrys* sp., *Harposporium* sp. Hence they participate in the microbiological balance in soil.

- Degradation of complex soil organic molecules like cellulose, lignin, hemicelluloses and pectin are done by fungi.

- They also function as plant pathogens and also act as biocontrol agents.

- Fungi participate in a unique symbiotic association with the roots of higher plants, the mycorrhiza which helps in solubilisation of major nutrients like phosphorus in the soil.

- Fungi also exist as a symbiont in the association of lichens which have great roles in the soil formation, soil texture and structure, cosmetic value, medicinal values and edible value.

- Soil fungi are commercially exploited for industrial enzymes like amylases, proteases, pectinases, etc.

- Soil fungi play a major role in manufacture of vitamin B complex and other vitamins. There are innumerable benefits of soil fungi still untold and unfolded.

Soil Algae

Algae are found in places with enough moisture and sunlight. Their number in soil is considerably less and ranges from 100–10,000 / gm of soil. They form a scum on the soil surface. Based on the colony morphology they are divided into unicellular, filamentous and colonial. Soil algae are divided into four classes:

Chlorophyceae   They contain chlorophyll in their chromatophores. They are called green algae, e.g., *Chlorella*, *Chlamydomonas*, etc.

Cyanophyceae   Apart from chlorophyll, they possess a blue pigment called phyco-cyanin. Hence they are called the blue-green algae, e.g. *Nostoc*, *Anabaena*, etc.

Xanthophyceae   They contain the yellow-green pigment called xanthophylls. Hence they are called the yellow-green algae, e.g, *Botrydium* sp.

Bacillariophyceae   They are unicellular/colonial which are surrounded by a highly silicified layer. They are golden brown in colour, e.g. *Navicula*.

Factors affecting algal population in soil

Organic matter   It has no appreciable bearing on algal distribution because of their photosynthetic metabolism. Large amount of organic matter in the soil does not favour the growth of algae since the bacteria and fungi outgrow them.

pH   Algae can grow at an optimum pH range of 7-10. They are often absent at pH <5 and uncommon at <6. Chlorophyta members appear in regions with a diversity of pH hence they dominate the algal flora of acid habitats because of the absence of other forms.

Moisture   Algal development is usually enhanced by increasing the supply of available water. Algal species increase during the rains and decrease during drought. Diatoms are the most sensitive to drying while the Chlorophyta and Cyanophycean members exhibit greater persistence in the form of resting stages for several years in the dry condition.

Season   Water availability is most favourable in spring and autumn when algae show maximum vigour. In the dried summer, the floral status is poor due to insufficient water and intense sunlight.

Depth   The need for sunlight is reflected clearly in the vertical distribution of the algae. Thus the population is most dense in the upper 5-10 cm and falls off with depth.

Inorganic nutrients   Algal members flourish at high levels of inorganic phosphorus and nitrogen (in the paddy fields).

Parasitic attacks   Bacteria, fungi and actinomycetes destroy the algal cells by producing extracellular enzymes.

Functions

- They play a major role in the flooded paddy fields by helping in nitrogen fixation in such an inhospitable habitat.
- One of the significance is its photoautotrophic nutrition, i.e. the generation of organic matter from inorganic substances. This is

particularly important in creating organic carbon *de novo* by colonising barren/eroded areas.

- Weathering rocks is another significant contribution of soil algae along with their unique symbiotic association to form the lichens.

- They contribute significantly to the soil structure and erosion control by binding together soil particles (due to the production of the hygroscopic exopolysaccharide on their outer cell surface).

- Some BGA can fix atmospheric nitrogen and thus have assisted in replacing nitrogen fertilisers to a great extent, e.g. *Nostoc*, *Anabaena*, *Tolypothrix*, *Chroococcus*, *Calothrix*, etc.

Soil Protozoa

All terrestrial species are microscopic with cells devoid of chlorophyll. They are not very important inhabitants of soil when compared to the already discussed members. Soil protozoa are classified on the basis of their means of locomotion as

Mastigophora   Move by means of long whip-like appendages called flagella, e.g. *Cercobodo*.

Sarcodina/rhizopoda/amoebae   They consist of a naked mass of protoplasm which move by pseudopodia, e.g. *Euglypha*.

Ciliata   They bear hairs on their body which help them to move, e.g. *Colpoda*.

A vast majority of them are saprobic feeders or are phagotrophic (direct engulfment of microbial cells) where they devour soil bacteria, e.g. *Enterobacter* sp., *Agrobacterium*, *Bacillus*, *E. coli*. When food is diminishing, the protozoa encyst themselves till the return of a favourable condition.

Factors affecting protozoan population in soil

Depth   Found in greatest abundance near the surface of the soil (in the upper 15 cm). They are scarce in the subsoils. The population is dense wherever bacteria are especially numerous.

Organic matter   Increase in organic matter favours growh of protozoa. Presence of large number of bacteria is shown an increase in organic matter.

Moisture level   Water is a major limitation to protozoan proliferation. Flagellates are tolerant to low moisture (found in Sahara desert). Ciliates are abundant only if moisture is high. Whenever water supply is low, the protozoans encyst themselves.

pH   Most protozoa exhibit no marked sensitivity to pH. They do not tolerate extreme acidity or alkalinity and fare poorly outside the pH range 6–8.

Temperature   The most favourable environment is cool and damp. Excessive warmth is detrimental.

Functions

- They serve to regulate the size of the bacterial community.
- Some are pathogens, e.g. *Entamoeba histolytica.*
- Some may help in the decomposition of plant remains.
- Protozoa may also enhance certain bacterial transformations such as utilisation of nitrogen or degradation of phosphorous containing organic materials.

Viruses

Soil viruses exist in the form of phages of bacteria, fungi, algae, actinomycetes and nematodes that live in soil. e.g. phages are found in *Rhizobium, Agrobacterium, Nitrobacter, Anabaena, Anacystis,* etc. Soil viruses are important disease-causing agents of many agronomic and horticultural crops. Some viruses like hepatitis virus can be discharged with the effluent of the septic tanks and move through the soil surface to the wells or other underground water supplies. Some viruses like TMV can remain buried for several years in the soil. Most of the viruses remain active in moist soil for a long time, Viruses can get adsorbed onto the clay particles or humus.

Functional role of viruses   They act as infectious agents of both plant and animal species including humans. They may play a role in the transmission of genetic material from one bacterium to another through transduction.

Review Questions

1. Define soil.
2. Define soil microbiology.
3. Describe the qualitative microflora of the soil.
4. Brief out the various extrinsic factors affecting the micorflora of soil.
5. Detail out the functions of soil microflora.

20

Soil Types and their Microflora

Except for organic soils whose constitution is different, the dominant substance in soil is the inorganic element silicon dioxide. There is an abundance of aluminium and iron. Calcium, magnesium, potassium, titanium, sodium, nitrogen, phosphorus and sulphur are in lesser amounts.

When compared to the element found in higher concentrations, it is the available nutrient in the particular soil that is of immediate significance to the microorganism. Elements found in lower concentrations than carbon and nitrogen may be present in amounts sufficient to satisfy their biological requirements.

A number of inorganic substances assimilated by microorganisms are anionic (bicarbonate, nitrate, sulphate, phosphates, etc.).

Soils are of different kinds:

Alluvial soils They are derived as silt depositions and their surface is reddish brown in colour. The texture of the soil is sandy loam and is deficient in humus, nitrogen and phosphorus. Microflora have an influence on the level of available potash. Certain bacteria are capable of decomposing aluminosilicate minerals and releasing a portion of potassium. *Bacillus* sp. and *Pseudomonas* sp. among bacteria and *Aspergillus* sp., *Mucor* sp., and *Penicillium* sp. among fungi are common. Since the soil is deficient in humus, one cannot expect high counts of heterotrophs which rely on organic matter for their energy.

Black soils The texture is silty loam with dark yellowish brown colour and is basic and sticky. The accumulation of gypsum is seen and the presence of humus, aluminium and iron is evident. It is rich in smectite and therefore possess high cation exchange capacity. It is also rich in iron, lime, magnesia, alumina and potash, but is deficient in nitrogen and phosphorus. The presence of iron oxidisers and reducers are evident in this soil, e.g. *Thiobacillus ferrooxidans* and *Metallogenium* sp. There may be presence of organic iron accumulators like *Pseudomonas, Bacillus, Serratia, Acinetobacter, Klebsiella, Corynebacterium, Nocardia, Streptomyces,* etc. A phenomenon possibly associated with microbial metabolism of iron is gleying by which such soil becomes sticky. In the gleyed sites, predominant aerobic iron reducers are *Bacillus* and *Pseudomonas* species.

Red soils They are red in colour and loamy or sandy. They are formed under a forest of broad=leaved trees in warm humid climates. They are acidic in nature (hence lot of fungal population is evident) and have a high concentration of organic matter. The microflora composed of bacteria, fungi and actinomycetes are varied both qualitatively and quantitatively.

Laterite soils They are non-sticky. Aluminium and iron are present whereas the soil is deficient in lime, magnesia, potash, nitrogen and phosphorus. Presence of iron oxidising and reducing organisms is common.

Forest and hill soils These soils are enriched by fallen leaves and are rich in organic matter and total nitrogen and in decomposers. Presence of nitrifiers, ammonifiers and denitrifiers is evident.

Desert soils These soils are sandy in nature and contain high amount of calcium carbonate and a low amount of organic matter. One can expect only high counts of chemolithotrophs like *Pseudomonas* species.

Saline and alkali soils In these soils, there is accumulation of salts of sodium, calcium and magnesium. Halophilic microflora are common.

Peaty and marshy soils Here, accumulation of large amount of organic matter (non-degraded) is seen. The soil is rich in carbon containing compounds due to non-decompostion of organic matter. Anaerobic microflora are present in huge numbers.

Review Questions

1. Trace out the different types of soil and their microflora.
2. What is gleying?

21

Quantification of Soil Microflora

Quantification is used to determine the amount or size of the soil particles. Soil is a highly heterogeneous ecosystem where conventional microbiological techniques only estimate a portion of the total number of bacteria. No one medium is nutritionally adequate for all the species present and the observed count represents only a fraction of the total.

Obtaining soil samples The soil samples are taken from a depth of 15 cm. 3–5 samples for each replicate are taken and mixed evenly. 10 –25 gm of soil is weighed and used for quantifying.

METHODS OF QUANTIFICATION

Soil Dilution and Plate Counts

This is the most widely method used for quantifying all types of soil microflora. To suppress bacterial growth, rose bengal or streptomycin can be added (to isolate fungi or actinomycetes). Thermophilic organisms are isolated by incubating at 55-60°C.

Media used The various media used for the various organisms are as follows:

- Nutrient agar for isolating soil bacteria.
- Rose bengal medium or potato dextrose agar with streptomycin for isolating fungi.
- Ken-knights medium, egg albumin medium or glycerol yeast extract agar for isolating actinomycetes.

- Benecke's medium for isolating algae.
- Skinner's medium for isolating protozoa.

How to isolate soil algae? The medium is filled in wide mouth bottles (up to 5 cm) and sterilized (20 lbs/20 min). One gram of soil is introduced. The set up is incubated at 30–35°C and lighted from above by 25 W lamps. After 15 days, algal growth is seen.

How to isolate soil protozoa? Take a spread plate containing a bacterial inoculum (*Enterobacter* or *E.coli*). Serially dilute the soil sample and inoculate it on the plate of nutrient agar containing the bacterial inoculum. After ten days of incubation, the presence of protozoa can be ascertained by microscopic examination for the cysts.

How to isolate phages? The soil is finely sieved and suspended in sterile distilled water (50 gm/150 ml). The suspension is filtered and collected in a sterile container. The filtrate is added to broth cultures of the host bacterium (*E.coli*). A pour plate is performed using nutrient agar with 1ml of the phase-bacterial suspension. After the plate is set, it is incubated at 27–38°C and examined at regular intervals. After 48–72 hours, patches of clear areas will be seen in the culture plates (plaques).

Buried Slide Technique for Direct Microscopy

This technique was devised by Rossi and Cholodny and is useful to study qualitative changes in soil microflora under the influence of soil amendments.

A clean glass slide is introduced into soil (for 1–3 weeks) at a depth of 10–15 cms. The top portion of slide is gently washed, air=dried and fixed over low flame. The slide is stained with erythrosine or rose bengal, dried and examined under microscope.

Direct Microscopic Examination of Soil

This method was devised by Conn in 1918. One gram of soil is mixed with 9 ml of 0.015% agar and shaken. One-tenth ml (0.1 ml) is transferred to the centre of a glass slide. The suspension is evenly distributed over a 4 sq. cm area and allowed to air dry. Then it is fixed in 0.1N HCl for a minute. It is then briefly immersed in water by placing in a water bath. The slide stained with rose bengal or erythrosine is examined under the microscope. A haemocytometer can be used to count the number of microorganism.

Indirect Estimation of Soil Microflora

When carbon-containing substrates are oxidised in soil, carbon dioxide is evolved which is generally taken as an index of the total activity of soil microflora.

Soil Enzyme Estimation

Total activity of soil microflora is the sum total of the number of activities of different microorganisms. For example, dehydrogenase activity can be a convenient laboratory technique for comparative studies.

Review Questions

1. List out the various media used for isolating soil microflora.
2. How is a direct microscopic examination of soil performed?
3. List out the methods for indirect estimation of soil microflora.

Methods of Studying Ecology of Soil Microflora

A wide variety of techniques can be used to evaluate the presence, types, and activities of microbes as populations, communities, and parts of ecosystems. A fundamental problem in studying microbes in nature is the inability to culture and characterise most organisms that can be observed. The quantification of soil microorganisms is very important to understand the ecological interactions between them and to increase the beneficial aspect of the soil interactions. The first step in the study of soil ecology is to collect the soil sample. It may require examination of samples at various levels including chemical, molecular, organisms as well as field scale ecosystems. Nevertheless, newer and more advanced techniques are rapidly changing the shape of soil microbiology, making the field an ever-challenging and exciting one to work in.

SAMPLE COLLECTION

Usually a soil corer is used for the sample collection from at least 1 to 2 feet depth which holds the core of collected soil. If microbial populations are presumed to be low in number in any soil, attractants or baits are used along with the soil corers to attract the microbes.

Sample Processing

The collected soil samples are processed for various kinds of studies. The study and the method of study is outlined and one of the methods is described below:

Determination of the form and arrangement of microorganisms in soil

1. Buried slide technique (Cholodny Rossi technique)
2. Impression technique

Direct examination of soil particles

1. Light microscopy
2. Electron microscopy
3. Cell counting by direct microscopy

Examination of root microbe interactions

1. The Fahraeus slide technique
2. Root observation boxes

Contact methods

1. Buried slide technique (Cholodny Rossi technique)
2. Impression tecnique

The Cholodny Rossi buried slide technique envisages two clean glass slides buried deep into the soil layers (usually 15¾25 cms deep) and remaining there for 2 to 3 hours. After that time, the slides are taken out without disturbing the soil and stained with eosin to observe the microorganisms directly under the microscope. The only disadvantage is that one cannot quantify the living microbes using this method. In another similar method, the glass slides are replaced with the electron microscope grids. After appropriate exposure in the soil, these grids are removed and observed under electron microscope. Another method envisages use of flattened glass capillary tubes (pedoscope) which are inserted into the soil for an appropriate time. These capillaries resemble the soil pore spaces and microbes enter them freely along with soil water. For further enrichment (to isolate specific group of organisms), the capillaries may be filled with nutrient solutions to attract them. For example, in soils rich in iron, counting of *Gallionella* (iron oxidising bacteria) can be performed easily by this method.

Isolation of Microorganisms

Direct methods of isolation

i. Direct isolation of fungi
ii. Direct isolation of bacteria
iii. Other direct methods of isolation

Indirect methods of isolation

    i.   Plating of untreated samples

    ii. Plating of pretreated sample (growth precluded)

    iii. Plating of pretreated samples (growth allowed)

Measurement of biomass and biovolume

    1. *Expression of results*

       a. Weight of soil constituent

       b. Volume of soil

       c. Specific surface area of soil or its constituents

       d. Surface area of ground

       e. Conversion of measurements to biomass

       f. Conversion of cell and spore counts

       g. Conversion of mycelial lengths

    2. *Chemical analysis of biomass / activity : Signal molecules*

       a. Chloroform fumigation extraction method

       b. Chloroform fumigation incubation

       c. Measurement of ATP: Luciferin reaction

       d. Lipid extraction : chloroform methane extraction

       e. FAME (Fatty acid methyl esters)

       f. Ergosterol: invasions of pathogenic fungi into plants

       g. LPS polymer measurement

       h. Nucleic acid analysis

       i. Direct lysis of DNA

       j. The polymerase chain reaction: Guanine and Cytosine (G+C) content

Expressions in the form of organism numbers (culture/growth based)

a. Plate counts

Disperse sample, dilute, apply to solid media in plates (composition determined by the investigator) incubate, count the colonies. Assume each colony is equal to one cell in the sample, effectively a colony forming unit (CFU).

Advantages

- Easy, only need plates and media (carbon source mixtures (glucose, yeast extract), protein mixtures (tryptone)).

- Versatile, media as specific or general. For specific media, use streptomycin to inhibit bacteria and selectively isolate fungi and use triphenylamine dyes to inhibit gram positive bacteria. Use xenobiotic compounds as carbon source to isolate biodegrading organisms.

Disadvantages

- No universal growth media is achievable, hence total population is never determined.

- Incomplete release of cells from particles results in underestimation of population.

- In cases where growth on specific compounds is being tested, contaminants in solidifying agents (i.e. agarose which is a mixture of polysaccharide polymers that forms a gel after heating and cooling) may be the source of interferences.

b. Liquid-based culture approaches (Most Probable Number [MPN])

This method can avoid solid media problems (contaminants, oxygen sensitivity) by sequentially diluting a soil sample in liquid growth media and observing the point at which growth is no longer detectable.

A 1 gm soil sample (dry weight) is suspended in 9 ml of media, this is a dilution of 10 (i.e. 1/10 or 10^{-1}). The sample is mixed, and 1 ml of the suspension added to 9 ml of fresh media. This is diluted further in order to give ten-fold dilution (overall 100-fold dilution is achieved). This procedure is repeated until the sample is diluted one million-fold (10^{-6}). The dilutions are incubated to allow growth of the organisms, and after a reasonable period these are examined for signs of growth or microbial activity. If growth or activity is detected at 10^{-5} and not at 10^{-6}, one could conclude that population was probably between 10^5–10^6 organisms per gram soil. To improve accuracy, results can be analysed by probability tables to refine the estimate and yield a *most probable number*.

MPN approach

- 10 g soil is diluted 10 fold (stock)
- Five 1-ml samples from stock are sequentially diluted 10X to an appropriate level (5 tubes per dilution; million fold = 6 levels × 5 tubes/level = 30 tubes).
- Tubes are incubated at 37°C for 24–48 hrs and observed for indications of growth/activity.
- A positive result means that at least one organism was added to the tube (i.e. cannot tell whether positive result originated with one cell or one million cells).
- Based on the number and pattern of positive tubes in a dilution series, the number of organisms in the original sample can be estimated from probability theory.

Advantages

- Enumeration of organisms difficult to grow on a standard solid medium culture can be achieved as for example, nitrifying bacteria which produce very little biomass or colonies.
- Anaerobic activity can be measured as in the case of denitrifying bacteria.
- Can enumerate plant symbionts as in the case of rhizobial enumeration by MPN using plants as detection systems (plant growth pouches replace tubes).

Disadvantages

- Limited by the precision of probability tables and labour intensive.
- Accuracy can be increased by increasing number of tubes per dilution and/or decreasing the dilution factor between levels. Either modification increases the labour input.
- Media bias, cell dispersal problems like other growth based techniques.

c. Epifluorescence microscopy

Uses fluorescent dyes that bind to components of microbial cells to provide contrast between cells and background.

Cells are separated from soil particles, diluted and samples are collected on a membrane filter (rather than on a plate as in the plate count approach), stained, observed and counted.

Acridine orange direct counts (AODC)

- Acridine orange binds to nucleic acids (DNA and RNA) cells may appear red-orange or green.
- Some have proposed that live against dead cells may be distinguished based on colour (live cells contain more RNA and appear red, dead cells lack RNA but may contain DNA and appear green.
- Thus AODC provides a measure of total number of cells irrespective of viability characteristics.

Advantages

- All observable cells can be counted

Disadvantages

- Inability to separate cells from abiotic debris (abiotic organic particles may fluoresce)
- Background fluorescence
- Cannot distinguish between living and dead cells
- Cannot distinguish between physiological groupings (e.g., organotrophs, denitrifiers, methanotrophs)
- Requires epifluoroscence microscope (very expensive)

d. Direct viability counts (DVC)

- For the variety of reasons outlined above, plate counts are justifiably considered to be inadequate for enumerating total viable bacteria in soil.
- To circumvent the problems inherent to culture techniques, a variety of non-culture based viability tests have been devised. The majority of these techniques utilise direct observation (microscopy) in combination with physical-chemical treatments.
- In principle, the response (or lack of response) elicited by the treatment can be visualised by microscopy and can be used to distinguish living against dead cells. In practice, all have shortcomings that make the results equivocal.

e. Fumigation incubation (FI)

In this method microbe carbon is made susceptible to mineralisation following exposure to chloroform vapours which leads to the release of carbon dioxide.

Chloroform exposure results in dissolution of microbial cell walls/ membranes making cellular components accessible to other microorganisms. The amount of carbon dioxide produced following fumigation is proportional to the amount of microbial biomass originally present.

` renders 99.9% of the biomass susceptible to mineralisation. Biomass mineralisation is not 100% because biomass decomposition is incomplete, and only a part of the biomass degraded is respired as carbon dioxide (the rest is incorporated into new cells). Soil organic matter (carbon) is not rendered ore degradable by fumigation.

Method

- The soil sample is divided into two portions (50 g each). One is fumigated, the other is not fumigated and used as a control sample to correct for "background" activity (i.e. CO_2 released from degradation)
- After fumigation, a 1% inoculum of the original "live" soil is added (e.g. 0.5 g mixed into 50 g). The control also receives this inoculum.
- The test and control samples are incubated for 7-10 days and total carbon dioxide production is measured. The latter is also referred to as the "flush".

Advantages

- Simple to set up and analyse.

Disadvantages

- Labour intensive
- Not applicable to low biomass samples (subsurface samples)
- Requires relatively large sample sizes
- Does not work well with acidic or calcareous soils

f. Use of ATP as a biomass indicator

- ATP extracted with a solvent from soil (0.5 g to 100 g)
- ATP measured by luciferin-luciferase assay:

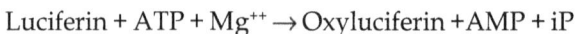

    Luciferin + ATP + Mg^{++} → Oxyluciferin + AMP + iP

    Light emissions quantified using a photometer and converted to ATP concentration by reference to a standard curve.

■ Correlate ATP to microbial biomass carbon

1 µg ATP = 120–500 µg of microbial biomass (250 µg microbe carbon /µg ATP)

Advantages

■ Specific for live cells

■ Direct measurement does not rely on indirect indicator (e.g. CO_2 evolution)

■ ATP measurement by the luciferin-luciferase assay is very sensitive with a detection limit as low as 50 fg = (10–15 g). Live cells contain 0.1 to 1 fg ATP.

Disadvantages

■ Cells' ATP content likely to vary depending on cell type and physiological status.

■ Extraction of ATP from cells may be incomplete and recovery from soil may be low.

■ Reagents are relatively expensive, specialised equipment needed.

g. Molecular techniques for analysis of soil microbial communities

Molecular techniques have recognised the limitations of culture-based approaches (lack of universal media for analysing microbial populations).

Molecular techniques circumvent these shortcomings by targeting molecules that serve as markers of a particular type of microbe, and which can be recovered or assayed without having to culture the microbes.

Many types of molecular techniques that have been applied to for the analysis of soil microbial populations including cell membrane components (fatty acids), cell surface components (proteins/carbohydrates exposed on the surface of the cell) and nucleic acids (DNA and RNA). The latter have been the focus of the greatest attention.

Approaches for utilisation of nucleic acid analysis

    a. *Types of information obtained from nucleic acid analysis*

■ Because DNA, mRNA and rRNA have different roles, their analysis provides different kinds of information about organisms and their activity.

- Genes (DNA) may be analysed by "probes" to provide information on the presence of a specific function in microbial populations.

- Determining amounts of specific mRNA present might allow activation (expression) of specific genes to be followed.

- rRNA analysis can allow identities of organisms in a population to be determined (the molecular clock aspect).

- It will focus on DNA and rRNA analysis. Recovery of mRNA from soil microbial populations has special problems that make it at present relatively uncommon.

b. *Recovery and use of nucleic acids to investigate soil microbial communities*

There are three phases in the molecular techniques applied to soil samples:

    i. nucleic acid extraction

    ii. extract purification

    iii. extract analysis

1. *Extraction and purification of nucleic acids*

- To extract DNA from soil organisms, their cells must be broken open or "lysed"

- Cell lysis can be achieved by treatment with cell wall degrading enzymes, chemicals that dissolve cell walls and membranes (phenol) or physical techniques (freeze/thaw, shearing). Combination of these methods are often used.

- Two strategies for lysing cells: one is the direct approach, in which cells are broken open directly in the soil. The other is an indirect procedure, where cells are removed from the soil prior to lysis. Of the two, direct extraction is more widely used because it is faster and generally gives higher nucleic acid yields.

- Neither method gives 100% yield of nucleic acids. In the indirect approach, yields are limited by the relatively low efficiency with which the cells are separated from soil. Nucleic acid yields from direct extraction are limited by incomplete lysis (degradation and reactions with soil constituents).

2. *Humic substances, interferences and removal*

- Humic substances (HS) are organic compounds formed by microbial degradation of plant materials.

- The exact structures are unknown, but these possess sufficient chemical similarity to nucleic acids that they are co-extracted along with nucleic acids. Problem with HS is that they may interfere with analysis techniques. A variety of approaches exist for removing HS, most widely used in column chromatography, which separates nucleic acids from HS based on molecular weight.

3. *Techniques for analysing nucleic acids extracted from soil*

   Two general approaches to examine the composition of nucleic acid mixtures are (i) through hybridisation to probes and (ii) determining the nucleotide sequence of specific nucleic acids in the mixture.

i. *Hybridisation to probes*

- A probe is a segment of DNA that is contained with in a gene that can be used to track a specific function ("function probes") or within rRNA and is thus useful for determining the presence of an organism in a community (phylogenetic probe).

- Probes are used to detect their targets by "hybridisation". There are three steps in this process: denaturation, hybridisation and detection. During denaturation, the probe and target DNA are converted by heating and/or chemical treatment from native double-stranded form to single-stranded molecule.

- The probe and target are then mixed, and because both DNAs are single stranded, the probe will be able to bind to ("hybridise") to its homologous sequence in the target.

- The probe-target hybrid can then be detected by a marker (incorporated into the probe when it was prepared) that emits a "signal" like radioactivity, colour or light.

- The amount of probe bound can be determined by quantifying signal intensity and thus allows the amount of target sequence present in the sample to be estimated.

ii. *Functional probes: Examples*

- Genes encoding enzymes that cause degradation of pollutants (pesticides, industrial chemicals) may be used to tack or detect pollutant degrading organisms in soil.

- Polychlorinated biphenyls (PCBs) are widespread soil pollutants. A number of PCB degrading bacteria have been isolated (cultured) and the genes encoding PCB degradation enzymes isolated (cloned).

- Layton *et al* (1994) showed a cloned PCB degradation could be used as a probe to monitor changes in PCB degraders, and that trends in gene probe results were consistent with those of cultural and chemical analysis of the soil. A major limitation was that the probe was relatively insensitive as the populations of PCB degraders apparently needed to increase by 4 orders of magnitude (i.e. 104 to 109 cells/g soil) before PCB degradation genes were detectable.

- Specificity of PCB gene probes can also be a potential problem for two reasons. First, PCB degradation genes are not all alike, so some PCB degraders may not be detected by the probe used resulting in "false negatives". Second, some PCB degraders have genes that are on the whole fairly similar, but have minor differences that result in major differences in the ability of the microbe to degrade PCBs. Thus, the probes do not reliably discriminate between the good, the bad (and the ugly) PCB degraders.

- The specificity of probes can also be advantageous when the goal is to track an organism introduced into soil. For example, genes that have novel, commercially useful functions maybe inserted into microbes (i.e. genetically engineered microbes "GEMs") and the organisms introduced into soil. For risk assessment, the fate of the novel genes can be tracked by using a gene probe. In this case, specificity is a benefit as it means there is a low probability of "false positives" (i.e., the probe hybridising to genes from the native soil microbes)

iii. *Phylogenetic probes*

- *Molecular clock aspect of rRNA:* different parts of the molecule undergo change at different rates. Some have remained essentially unchanged for millions of years, and are thus the same in all organisms, these are the conserved regions. Other regions are variable, with the nucleotide sequence changing at a variety of rates.

- Conserved regions can be targeted by "universal" probes, that will detect any organism. Probes to variable areas may distinguish between organisms at levels ranging from the domain to species level. For example, analysis of microbial populations in sewage sludge using phylogenetic probes. In this case the objective is to determine how the in situ population composition (i.e. in the sludge itself) compares to that recovered by plates.

- Another approach for determining the phylogenetic identities of soil organisms is to analyse the nucleotide sequence rRNAs extracted from soil.

- These sequences can then be compared to those of other organisms via computer databases to determine similarities.

- Results of these comparisons are analysed by computer algorithms to generate clusters, and the output displayed in 'dendrograms', in which individuals are sorted based on percent similarity. Those that are identical, (100% similarity) are clustered together. Organis ms with lower degrees of similarity are divided in separate clusters.

Advantage

- By definition they are representative of an organism and so results are less susceptible to false negatives. Like function probes, the results can be made quantitative.

Disadvantage

- In most cases, these do not directly provide information on function.

Hence, any one method based on the various above-mentioned methods to analyse the soil microflora can be followed in order to study the ecology of the same.

Review Questions

1. Describe in detail the various methodologies for studying ecology of soil microflora.

23

Role of Microbes in Soil Fertility

S oil fertility is the capacity to supply proper amounts of different nutrients in the proper proportion for the growth of crops. The availability of both inorganic and organic matter determines the soil fertility. The inorganic matter of soil comprises all the essential and trace minerals present in the soil in the form of salts (acidic and basic). The inorganic element either gets adsorbed onto the clay particles or get dissolved in the soil water.

The organic matter in the soil exists mainly as humus or as partially decomposed (plant and animal tissues). It is also prepared artificially as farm yard manure, green manure/green leaf manure, compost, vermicompost, biofertilizers, etc.

On the whole, the balanced availability of both inorganic and organic matter in the soil determines the soil fertility. Indirectly, these organic and inorganic matter help in the proliferation of various qualitative microflora that play a very vital role in maintaining the nutritional balance of the soil. Thus microorganisms have a great role to play in determining a soil's fertility, for without a proper distribution of microflora, no soil can support plant growth which speaks of its fertility.

Microorganisms in soil affect the fertility of soil by means of physical or chemical changes.

HOW DOES MICROORGANISM AFFECT THE PHYSICAL STATUS OF SOIL

- It is the microflora that are responsible for the formation of soil from barren rocks due to the collective activity of algae, moss and lichens that colonise the bare rocks, produce organic acids which dissolve the primary minerals and release the nutrients contained in them for plant growth.

■ Microflora improve soil structure by improving the soil texture, i.e. by making the soil more loamy. For example, algae and some bacteria that have exopolysaccharide secretion onto their cell surface due to their hygroscopic properties, bind more water molecules to their surface.

■ Presence of these microbes in a sandy soil, converts the soil to more loamy by binding more soil particles onto their surface (by increasing the moisture). Such soils are improved in their mineral binding capacity and thus their fertility.

■ Presence of soil microflora allow a lot of gaseous exchange and thus favours better soil aeration.

■ Presence of microfauna like worms or earthworms increase soil aeration and allows better nutrient absorption by the plant roots.

■ Presence of soil microflora tend to maintain the balance of pH in the soil by excreting metabolites (acidic and basic) in order to facilitate better absorption of mineral nutrients by the plants.

■ The microflora present in the soil prevent soil erosion thus maintaining the fertility of the soil.

ROLE OF MICROORGANISMS IN MAINTAINING THE CHEMICAL BALANCE IN THE SOIL

Microorganisms accomplish this by taking part in all the major element transformations (mineralisation and immobilisation) in soil. They convert the complex organic nutrients into simpler inorganic compounds (mineralisation) so that plants can make use of the nutrients and these microbes absorb the simpler minerals (immobilisation) and prevent them from leaching out and thus conserve the essential nutrients in the soil so that when they die and become a part of the organic matter, these essential nutrients are once again mineralised by the microorganisms for plant use. Thus the soil fertility that is created by the microbes is also conserved by the same.

The major transformations by the microbes are:

■ carbon

■ nitrogen

- phosphorus
- sulphur
- iron
- manganese
- potassium
- trace elements

Apart from the soil-based transformations, bacteria like *Azotobacter*, *Azospirillum* are able to convert the atmospheric gaseous nitrogen to reduced inorganic nitrogen for direct assimilation by plant roots thus limiting the use of chemical nitrogenous *fertilisers*.

Besides, some microbes like *Rhizobium, Bradyrhizobium*, etc. form an explicit symbiotic association with the roots of leguminous plants where they form root nodules and fix atmospheric nitrogen directly in the nodule.

MICROORGANISMS AND FARMYARD MANURE

Material released from the decompostion of the mixture of animal dung and urine soaked litter by microorganisms is called farmyard manure (FYM).

Application of FYM, green manure and compost greatly increase the soil fertility by releasing the nutrients that have been mineralised by the microflora contained in them.

Thus microbes have a very significant role to play in increasing as well as maintaining soil fertility.

Preparation of FYM

A manure pit with roof (0.9 m deep, 2.4 m wide and 5 m long) is dug out. A mixture of cow dung, urine-soaked litter is evenly spread at the bottom. This is continued till the heap rises 30 cm above ground. Decomposition of the organic manure by the microbes occurs and FYM is ready for use after 6 months. The composition of FYM is 0.32% N, 0.05% P, 0.25% K, 1.20% Ca and 0.33% Mg. When applied to soil, it improves soil structure and soil fertility.

Functions of Essential Nutrients

- **CHO** Major constituents of all organic compounds.
- **N** Major structural constituent of cell protein, nucleic acids, chlorophyll, enzymes, ADP, ATP, hormones, etc.

- **P**   Constituent of cell nucleus, nucleic acids, ADP, ATP, essential for cell division.

- **K**   For formation of amino acids and proteins, transfer of carbohydrates and proteins from leaves to roots. Uptake of essential elements like N, P, Ca. Regulates permeability of cell membrane. Increases hydration of protoplasm and activates a number of enzymes. It also increases resistance of crops to unfavourable conditions.

- **Ca**   Regulates permeability of protoplasm. It is the structural component of chromosomes.

- **Mg**   Constituent of chromosomes and polyribosomes and chlorophyll.

- **Fe**   Strutural part of enzymes and their activators, cytochromes, ferredoxin and haemoglobin.

- **Mn**   Constituent of nitrite reductase and activates enzymes.

- **Cu**   Constituent of oxidation and reduction enzymes.

- **Zn**   Constituent of enzymes.

- **Mo**   constituent of enzyme nitrate reductase. Essential for nitrogen fixation by free-living bacteria by *Azotobacter* species.

- **B**   Needed for proper development of vascular elements.

- **Cl, Co and Na**   Co is the structural component of vitamin B12. Na maintains osmotic balance of cells. Cl is involved in oxygen synthesis during photosynthesis.

TESTS FOR SOIL FERTILITY

Soil nutrients are gradually withdrawn from the soil by the crops growing on it. The available nutrient content of soils is determined by soil tests. This can be done analytically as well as microbiologically.

Available Nutrient

That form of each soil nutrient most responsible for increasing crop yield is called the available form of nutrient, e.g. available form of phosphorus is phosphate, available form of sulphur is sulphate and available form of nitrogen is ammonia or nitrates. Testing or isolating the presence of the particular microorganism responsible for the transformation of a particular nutrient (e.g., presence of *Nitrobacter* in the soil indicates the transformation of nitrogen in the form of available nitrate) is the basis of microbiological testing of soil fertility.

Collection of Soil Sample

Soil is collected from a depth of 20 cm and a minimum of 10 samples collected and pooled can be used for the tests.

Testing the Presence of Available Nitrogen

Test for ammonifiers

Medium used   Ammonifying medium.

In this medium, protein source is provided by the peptone. If ammonifiers are present in the soil sample, the protein nitrogen is converted to inorganic ammoniacal nitrogen which is chemically analysed by Nessler's reagent.

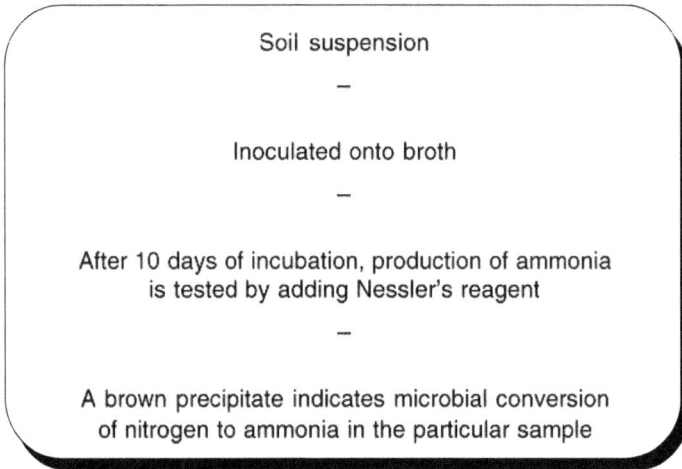

Soil suspension

–

Inoculated onto broth

–

After 10 days of incubation, production of ammonia
is tested by adding Nessler's reagent

–

A brown precipitate indicates microbial conversion
of nitrogen to ammonia in the particular sample

Test for nitrifiers

Medium used   Ammonium calcium carbonate medium.

Here the source of nitrogen is in the ammoniacal form. When the soil sample is inoculated onto the broth, if nitrifiers are present, they convert the ammoniacal nitrogen to nitrites (*Nitrosomonas*) and nitrates (*Nitrobacter*).

```
Soil suspension
        –
Inoculated onto broth
        –
Incubation at 28°C for 3 weeks
        –
The broth solution is mixed with Griess
        Ilosway reagent (3 drops)
        –
Development of reddish purple colour
indicates the presence of nitrite (Nitrosomonas)
        –
Failure of colour development indicates the presence
    of Nitrobacter, i.e. conversion of nitrite to nitrate
has taken place (since this conversion is very rapid)
```

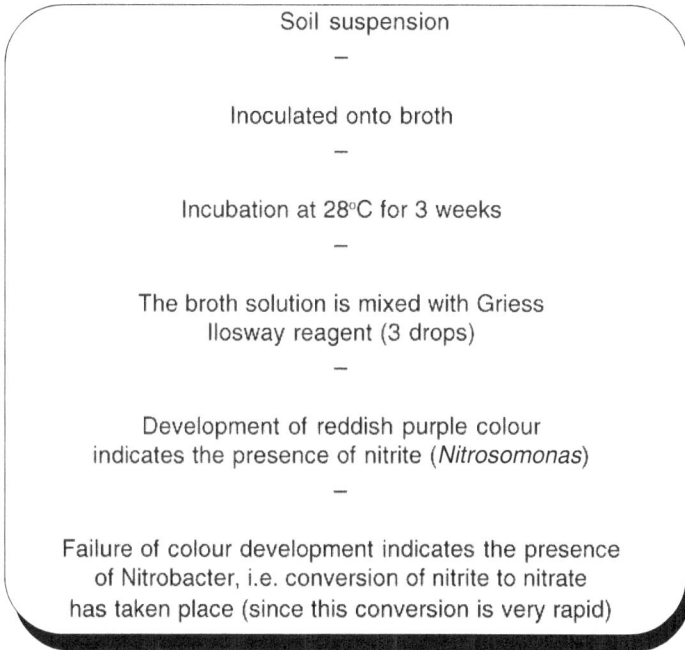

In any case this nitrifying process can be determined by one of the above-mentioned methods. All these methods indicate the microbiological role of maintaining the soil fertility, i.e. soil nitrogen. These methods also indicate the requirement of fertilizers in the particular soil.

Testing the Presence of Available Phosphorus

Soil phosphorus is usually in the unavailable form (mainly the inorganic phosphorus which occurs in the insoluble form of triphosphates of calcium or magnesium). Organic phosphorus comes from phytin, animal and plant tissues, bones, etc. The unavailable form of phosphorus is made available by the range of microorganisms:

Bacteria   *Pseudomonas, Bacillus.*

Fungi   *Aspergillus niger, Aspergillus flavus, Penicillium* sp., *Fusarium* sp., *Cladosporium* sp.

Medium used   Pikovskaya's medium.

Phosphorus is presented by tricalcium phosphate (insoluble form of phosphate).

Soil suspension (1ml)

–

Pour plated with the medium

–

Incubated at 37°C for 48–72 hours

–

Clearing around bacterial and fungal growth
indicates phosphorus solubilisation.

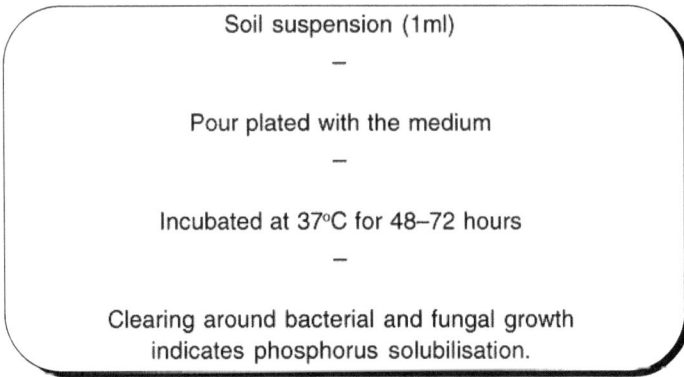

Microorganisms produce organic acids like oxalic acid, succinic acid, acetic acid and citric acid that solubilise the insoluble phosphorus in tricalcium phosphate and converts it to soluble phosphate precipitating the calcium thus making the inorganic phosphorus as phosphate available to the plant roots.

Test for Presence of Free-living Nitrogen Fixers (*Azotobacter*) in the Soil

Medium used   Ashby's mannitol agar or Jensen's medium.

A trace of molybdenum is used in the medium which plays a vital role in the enzyme system used for fixing nitrogen. Mixing the soil sample in the medium and plating gives rise to big translucent colonies of *Azotobacter*. Key feature of this test is that the medium is deficient in nitrogen source.

Review Questions

1. Define soil fertility.

2. Discuss the role of soil microflora in maintenance of physical and chemical structure of soil.

3. Brief out on certain tests for soil fertility.

PART V

MICROBES AND HYDROSPHERE

Microbiology of Water

Bacteriological Analysis of Water

Water Pollution

Eutrophication

Waterborne Diseases

Purification of Water

Recycling of Water

24

Microbiology of Water

Water is indispensable for life. The basic human physiological requirement for water is about 2.5 litres per day. This drinking water would be free from chemical as well as microbial contaminants, since the potential of contaminated water to transmit disease is very high. Potable waters are normally tested for their quality based on various factors because water quality is very important in both health and industrial aspects of economy. The most frequently used indicator organism is the normally nonpathogenic coliform bacterium, *Escherichia coli*.

The great solvent power of water makes the creation of absolutely pure water a theoretical rather than a practical goal. The problem, therefore, is one of determining what quality of water is needed to meet a given purpose and then finding practical means of achieving that quality. The problem is further compounded because every use to which water is put¾ washing, irrigation, flushing away wastes, cooling or making paper¾ adds something to the water. The term water pollution is referred to any type of aquatic contamination between the following two extremes: (1) a highly enriched, over-productive biotic community, such as a river or lake with nutrients from sewage or fertiliser (cultural eutrophication) and (2) a body of water poisoned by toxic chemicals which eliminate living organisms or even exclude all forms of life.

A vague understanding of the need to protect water systems that are used for drinking, from contamination with waste and wastewater is documented in historic documents, for instance in the Bible. The regular outbreaks of diseases like typhoid fever and cholera were thought to be related not to water, but to local atmospheric conditions. Despite this misleading theory, John Snow concluded from epidemiological evidence that a drinking water pump was the cause of cholera outbreak in London. By removing the handle of the pump in 1854, he was able to stop the cholera outbreak.

The necessity of resource protection and drinking water treatment became evident when the connection between bacteria in drinking water and the outbreak of various diseases was made. Today, in most industrialised countries, drinking water is ranked as food, and high standards are set for its quality and safety. The strict requirements for microbiological factors specify that bacteria content should be very low and that no pathogenic microbes should be detectable.

The fear of classic waterborne infectious diseases like cholera and typhoid fever have been lost in developed countries.

Water may contain various contaminants and pollutants and it is extremely difficult to obtain pure, potable water. Faecal contamination of drinking water supplies is a potential problem. Presence of disease causing pathogenic microbes is also common. In addition to microbial contaminants, organic and inorganic colloids are also present in water. Clays, microbial debris, and reduced iron and manganese compounds are also common. When water is obtained from open surface water bodies, floating or large suspended solids such as leaves and branches may be present. Hence, to render water fit for drinking and domestic use, treatment of the water is necessary. The purpose of water treatment is to convert raw water into drinking water suitable for domestic use. Most important is the removal of pathogenic organisms and toxic substances like heavy metals which cause serious health problems. Other substances to be removed include suspended matter causing turbidity, iron and manganese compounds imparting a bitter taste and excessive carbon dioxide corroding concrete and metal parts.

Drinking water is obtained from different sources, and various impurities ranging from branches of trees to invisible microbes may occur in these waters and have to be removed before the water is supplied to the public. To render the water safe for drinking, it has to be treated properly.

SOURCES OF DRINKING WATER

Water is the most common and important chemical compound on earth. The availability of drinking water has been the most critical factor for survival throughout the development of all life. As the population increased, the natural supply of water became limited, sophisticated techniques and systems were developed to obtain access to new water reservoirs and to distribute water for irrigating and drinking. Besides hygienic problems caused by unsanitary wastes, the rapid development of industry, especially the development of the chemical industry, has resulted in an ever-present contamination of all kinds of natural water systems.

Surface water Streams, rivers and lakes are the major sources. These waters originate from ground water and rain water (surface run off). Surface run off contributes to turbidity and microorganisms.

Ground water Wells and springs originating from infiltrated rain water which flows through the underground are also common sources. A little contamination of the ground water occurs from organic and inorganic soil particles, animal and plant debris, fertilisers, pesticides, microbes, etc. as it flows through the soil layers. Partial removal of microorganisms occurs by the death of cells due to lack of nutrients.

Rain water This is of high quality since the only possible source of contamination is airborne microorganisms that too in very low numbers.

Based on quality, water can be of the following types.

Potable water Clean, safe water free from disagreeable taste, odour, harmful chemicals, turbidity and microorganisms is called potable water.

Polluted water Water with added substances which impart colour, odour and taste is polluted water.

Contaminated water This is water which is unsafe for drinking since it may have added discharges from human or animal intestines, or is rendered dangerous by addition of poisonous chemicals.

Water may contain various contaminants and pollutants including faecal contamination, presence of disease causing pathogenic organisms, organic and inorganic colloids, clay, microbial debris, reduced iron and manganese compounds and floating or large suspended solids (leaves or branches). To render water fit for drinking and domestic use, treatment of water is necessary. It is essential that the water used for drinking purpose is periodically examined for microbiological parameters.

The detection and estimation of pathogenic bacteria is a tedious work as the number of these organisms in water is very low. The commonly associated microorganisms in potable water are *Salmonella typhi, Vibrio cholerae, Salmonella paratyphi*, other enterococci like faecal streptococci, *Streptococcus faecalis, Clostridium perfringes* and *Bifidobacterium* species, *Salmonella typhimurium* and *Shigella dysenteriae*.

The various methods used in the microbiological examination of potable water include:

- *Determination of standard plate count* This provides density of aerobic and facultative bacteria which can grow at 37°C in water sample.

- *Most probable number* This method statistically signifies the probable number of coliforms or other contaminating microorganisms that may be present in water.

- *Filtration* By filtering a known volume of water to be analysed through a membrane filter apparatus which has a provision for filtering the microorganisms on its filter pad (having a pore diameter of 0.45 m), one can count the number of microorganisms by plating the filter pad directly on the medium.

- *Gram staining* It gives a direct report about the presence of microorganisms in water but it is not reliable like the previously mentioned methods.

Review Questions

1. Brief out the sources of drinking water.

2. What is the difference between polluted water and contaminated water.

3. Define potable water.

25

Bacteriological
Analysis of Water

The first step in the bacteriological analysis is sampling. Samples of water or sewage for bacterial analysis should be collected in suitable bottles that have been carefully cleaned, rinsed in clean water (preferably distilled water) and sterilised.

Two kinds of bottles may be used: (1) clean sterile bottles and (2) clean sterile sodium thiosulphate-treated bottles. Water or sewage effluents containing residual chlorine should always be collected in sodium thiosulfate-treated bottles. All other samples may be colleted in clean sterile bottles. In collecting samples, extreme care should be exercised to avoid contaminating parts of the bottle coming in contact with the water. The stopper should be handled without removal of the protective cover. Bottles should be filled to three-quarters of their capacity. If samples are collected in sodium thiosulphate-treated bottles, care must be exercised not to rinse the bottle and lose the sodium thiosulphate.

All samples should be tested as soon as possible. In warm weather, if the transportation period exceeds one hour, the sample should be iced. The time for transportation and storage should not exceed six hours for impure waters and 12 hours for relatively pure waters. Samples should be stored at a temperature between 6° and 10°C.

PURPOSE OF BACTERIOLOGICAL ANALYSIS

The main purpose of bacteriological analysis of water is to determine the potability of water. There are three groups of bacteria present in abundance in the intestinal tract of man and animals¾the coliforms, the anaerobic lactose-fermenting sporeformers and the faecal streptococci. Of these three groups, the coliform organisms are more closely related to the intestinal pathogens (typhoid, dysentery and paratyphoid organisms) and hence are affected by storage, sedimentation, chlorination and other natural or induced processes of purification to approximately the same degree. Positive tests for *E.coli* do not prove the presence of enteropathogenic organisms but do establish this possibility. Because *E. coli* is more numerous and easier to grow than the enteropathogens, the test has a built-in safety factor for detecting potentially dangerous faecal contamination. *E.coli* meets many of the criteria for an ideal indicator organism but there are limitations to its use as such and various other species have been proposed as additional or replacement indicators of water safety.

Thus the absence of the coliforms would indicate also the absence of the intestinal pathogens. The test for the coliform group thus is an indirect one. The test should be quantitative to measure the density of the organisms in water.

COLIFORM GROUP

The coliform group includes two genera of bacteria, the *Escherichia* and the *Aerobacter*. This group is defined as 'all aerobic and facultative gram negative non-spore-forming bacilli which ferment lactose with gas production'. The identification of the group can be limited to the following characteristics:

1. (a) ability to ferment lactose with gas formation in nutrient lactose broth followed by

   (b) ability to ferment lactose in brilliant green bile lactose broth

or

2. (a) ability to ferment lactose with gas formation in nutrient lactose broth followed by

   (b) ability to grow on surface of eosin-methylene blue agar. The organisms isolated from these mediums should ferment lactose in nutrient lactose broth. The organism should be gram negative and not form spores.

For *E.coli* to be a useful indicator organism of faecal pollution, it must be differentiated readily from nonfaecal bacteria. The conventional test for the detection of faecal contamination involves a three-stage test procedure: the presumptive test, the confirmed test and the completed test.

Presumptive Test

The first step in the bacteriological analysis is the development of the coliform organisms in standard lactose broth. The coliform group ferments lactose with the production of gas. The production of gas from lactose is a presumptive test for the group but not a definitely positive reaction because a few other bacteria, not necessarily of sanitary significance, may also produce the same reaction. The most common bacteria other than the coliform group which produce gas are the sporeformers, both anaerobic and aerobic.

The amount of gas produced in a 48 hour period of incubation at 37°C has no particular significance. Some strains of the coliform organisms produce only slight amounts of gas. Any amount of gas should be considered a positive presumptive test if the gas appears within 48 hour of incubation at 37°C.

Confirmed Test

To demonstrate the aerobic nature of the coliforms, smears are made on EMB medium from the tubes showing gas. The smears should be made such that discrete colonies are produced. It must be remembered that the growth in the lactose broth is due to many species of bacteria in addition to the coliform organisms. Unless discrete colonies are formed, it is impossible to make isolations representing pure cultures because the confluent growth is a mixed population. The appearance of the confluent growth as indicating coliform organisms may be very misleading owing to the multiplicity of reactions induced by the different species of bacteria that may be present.

By smearing the surface of the agar plate, the coliform group is able to grow but the anaerobic sporeformers are eliminated as they are unable to grow aerobically. The dyes in the medium also act as a deterrent to many of the undesirable bacteria. To distinguish the coliform organisms from other organisms that might grow on the agar plate, the use of special media like EMB and Endo agar give the coliform colonies a distinctive appearance that aids in their identification.

Completed Test

To make sure that either typical or atypical colonies are members of the coliform group, the colonies fished are planted into lactose broth and on an agar slant. If the organisms produce gas in the fermentation tube, then the corresponding agar-slant culture should be examined for spores and gram's reaction. These tests eliminate the aerobic, lactose-fermenting, gram positive sporeformers. The only organisms that successfully carry through these tests are the members of the coliform group.

A rapid test is also done for identifying the coliforms. In this procedure the presumptive test is made as usual but instead of smearing eosin-methylene blue agar or endo's medium, plantings are made into one of the following mediums: brilliant green bile lactose broth or crystal violet broth. If possible, platings should be made from the presumptive lactose-broth tubes as soon as gas appears. The plantings are incubated for 48 hours at 37°C. If gas appears during this period of incubation, the test is reported as positive.

E.coli meets many of the criteria for an ideal indicatory organism but there are limitations to its use as such and various other species have been proposed as additional or replacement indicators of water safety.

Review Questions

1. What is the purpose of bacteriological analysis of water?
2. What do you mean by the term 'coliform group'?
3. Describe the three-stage test procedure for detection of faecal contamination.

26

Water Pollution

Polluted water is one which consists of undesirable substances rendering it unfit for drinking and domestic use. Even the highest quality distilled water contains dissolved gases and to a slight degree, solids. Polluted waters need not be contaminated. The great solvent power of water makes the creation of absolutely pure water a theoretical matter than a practical goal. The different types of pollutants and their effects are given in Table 26.1.

SOURCES OF WATER POLLUTION

Organic forms It includes all living forms suspended in water. For example, bacteria, algae, aquatic forms and decaying matter.

Synthetic organic chemicals It includes the herbicides that are used to kill weeds which may percolate through soil and get dissolved in ground water. Pesticides, fungicides which are used to kill insect pests and disease microbes may contaminate water through soil percolation or by aerial spray system.

Inorganic forms It includes salts of calcium, magnesium and sodium in the form of carbonates, sulphates, chlorides, nitrates. These salts impart bad taste, hardness, alkalinity, etc. Iron oxide and manganese impart red, black or brown colour to the water. Radioactive wastes from atomic reactors are excessively harmful to aquatic life if released into water systems.

Sewage Rivers and wells are the main source of drinking water. These sources are usually polluted by domestic sewage, human and animal excreta. Sewage may also percolate through soil and contaminate potable waters. Sewage may be the main cause of waterborne diseases as they contain pathogenic microorganisms.

Table 26.1 *Pollutants of water and their effects*

Pollutants	Effects
Organic wastes: sewage, decaying plants, animal manures, wastes from food processing plants, oil refineries andleather paper and textile plants.	Increase BOD of water
Pathogens	Cause diseases in humans who drink the water
Inorganic chemicals and minerals	Increase the salinity and acidity of water and render it toxic.
Synthetic organic chemicals herbicides, pesticides, detergents, plastics, industrial wastes.	Cause birth defects, cancer, neurological damage and other illness.
Plant nutrients	Cause uncontrolled growth of aquatic plants (eutrophication) impart undesirable odour and taste to drinking water.
Sediments from land erosion	Cause silting of water ways and destruction of hydroelectric equipments near dams. Reduce light reaching plants in water and oxygen content of water.
Radioactive wastes	Cause cancer, birth defects, radiation sickness.
Heated water	Reduces oxygen solubility in water, alters habitats and kind of organisms present, encourages growth of aquatic life but can decrease the growth of desired fishes.

Industrial effluents Untreated industrial effluents can pollute water sources. In Japan, people were poisoned either directly or through eating fish contaminated by traces of mercury from a metallurgical plant. Even in many parts of India ground water remains polluted with toxic chemicals like arsenic which play havoc in the lives of the people living in that part of the country (West Bengal). Wastes from paper and pulp mills contain a heavy load of fibrous material that usually block pipes and sewers. Effluents from plastic factories contain a high content of phenols that give obnoxious taste to the water. Industrial wastes from metal finishing and electroplating units contain heavy metals and cyanides that add to water pollution.

Faecal pollution Human and animal excreta may enter into potable water sources either through sewage infiltration or due to lack of sanitation. Pathogenic microorganisms thus enter water sources and are responsible for major outbreaks within communities.

Temperature Water used for cooling atomic reactors and power stations are discharged into rivers and streams. It increases the temperature of water and reduces oxygen content.

Oils Both sea and river waters are polluted by crude oil due to accident, transport, etc. Seawater may also be polluted due to offshore oil production and oil refineries which discharge their wastes into marine system.

Harmful effects of oil

- It causes fire in sea and river as a result of which aquatic plants and animals are completely destroyed.
- It prevents oxygenation of water.
- Oil deposits in animal systems like gills and scales of fish results in death of the animal.
- It is estimated that presence of oil in water is a serious obstacle to the process of carbon assimilation in the phytoplankton.

ECOLOGICAL EFFECTS OF WATER POLLUTION

Minamata disease This disease is caused by mercury poisoning. The disease first occurred in Minamata, a small town in Japan and the primary cause was mercury which was released as a byproduct of the plastic industry. Mercury poisoning occurred in marine animals and this led to death of birds, cats and dogs which ate marine animals like fish, crabs and shell fish. Initial symptoms of Minimata disease include numbness of lips, limbs and tongue. Impairment of motor nerve system, deafness and blurring of vision. The disease then progresses to brain dysfunction.

Mortality of plankton and fish　Chlorine which is added to water to control growth of algae and bacteria may persist in streams to cause mortality of plankton and fishes.

Reduced productivity　Intensive agriculture increases the amount of silt in lakes and rivers. Silts prevent penetration of light and thus reduces primary production. Siltation is a phenomenon by which gills of fishes are deposited with silt. This causes heavy mortality among fishes.

Red tide　When coastal waters are enriched with nutrients of sewage, dinoflagellates multiply rapidly and form a bloom. This is referred to as 'Red tide'. These blooming dinoflagellates liberate toxic metabolic byproducts which result in large scale death of marine fishes.

Biological oxygen demand　Sewage enriches water systems with nutrients. This causes rapid growth of plankton and algae which leads to oxygen depletion in water. Biological oxygen demand (BOD) is the amount of oxygen required by microorganisms in water. BOD is higher in polluted water. Increased BOD lowers the content of dissolved oxygen in water causing suffocation and death of aquatic flora and fauna.

Methaemoglobinema　Nitrates used in fertilisers enter the intestine of man through drinking water, where it is converted into nitrites. Nitrite is absorbed in the blood where it combines with haemoglobin to form methaemoglobin. Methaemoglobin cannot transport oxygen. This leads to suffocation and breathing trouble especially in infants. This disease is called methaemoglobinema.

RECENT ASPECTS OF WATER POLLUTION

Eutrophication　Domestic sewage and fertilisers add large quantity of nutrients such as nitrates and phosphates to the fresh water systems. The rich supply of nutrients favour growth of blue-green algae, phytoplanktons and green algae. The increased productivity of lakes and ponds brought about by nutrient enrichment is known as eutrophication. As the algae use oxygen of water for respiration, oxygen depletion occurs. Nutrient depletion also occurs leading to death of algae and other phytoplanktons. This in turn leads to the death of zooplanktons and fishes. Thus eutrophication leads to complete depletion of living forms in an ecosystem.

Biofilm formation　Biofilm or microbial film is caused by adhesion of bacteria to surfaces. Biofilm consists of accumulation of cells, extracellular products, inorganic and organic debris.

Microorganisms in biofilm include:

- *Gallionella*   Iron oxidising bacterium
- *Pedomicrobium manganicum*   Manganese oxidising bacterium
- *Flavobacterium, Serratia, Chromobacterium*   Pigmented bacteria
- *Pseudomonas, Clostridium, Legionella, Campylobacter*   Pathogenic bacteria
- *Acanthamoeba*   Microorganisms causing chronic eye infection
- *Naegleria*   Meningoencephalitis

Bioremediation   It is the process by which microorganisms convert toxic or ecologically harmful materials into harmless molecules. The toxic organic compounds are used as carbon and energy source by microorganisms. But inorganic compounds are accumulated intracellularly.

Bioaugmentation   Many low molecular weight compounds are not metabolised efficiently by natural microorganisms. Such compounds are known as recalcitrant chemicals. Recalcitrant chemicals include solvents, pesticides, herbicides etc. Some microbes are capable of metabolising these compounds. The introduction of microbes into a contaminated environment to bring out detoxification is termed as bioaugmentation.. It is used in detoxification of halogenated compounds like polychlorinated biphenyls (PCBs). Organisms like *Micrococcus*, *Corynebacterium*, *Nocardia* and *Penicillium* are capable of hydrocarbon degradation. They are used to degrade oil in industrial wastes, oil spills in oceans and in oil-soaked soil near refineries.

Review Questions

1.   Tabulate the various water pollutants and their effects.
2.   Describe in detail the various sources of water pollution.
3.   List out the harmful effects of oil as a water pollutant.
4.   Trace the ecological effects of water pollution.
5.   What is eutrophication?
6.   Give a few examples of microbial biofilm.
7.   What is bioremediation?
8.   What is bioaugmentation?
9.   What are recalcitrants?

27

Eutrophication

According to Hutchinson (1969), eutrophication is a natural process which literally means 'well-nourished or enriched'. It is a natural state in many lakes and ponds which have a rich supply of nutrients, and it also occurs as a part of the aging process in lakes, as nutrients accumulate through natural succession.

Eutrophication becomes excessive, however when abnormally high amounts of nutrients from sewage, fertiliser, animal wastes and detergents, enter streams and lakes causing excessive growth or 'bloom' of microorganisms and aquatic vegetation. Most secondary sewage treatment plants, though, precipitate solids and inactivate most bacteria in domestic sewage, yet they do not remove the basic nutrients such as ammonia, nitrogen, nitrates, nitrites and phosphates. These nutrients stimulate algal growth and lead to *plankton blooms*. Some plankton blooms, particularly those of blue-green algae produce obnoxious odours and tastes in waters. Others, such as the dinoflagellate blooms or 'the red tide' of southern coastal regions, produce toxic metabolic products which can result in major fish kills. Plankton blooms of green algae do not always produce undesirable odours or toxic product, but still create problems of oxygen supply in the water. While these blooms exist under abundant sunlight, they contribute oxygen to water through photosynthesis, but under conditions of prolonged cloudiness, they begin to decay and consume more oxygen and with heavy load the oxygen content of the water may diminish below the point where most fish cannot survive.

As the conditions in the water become anaerobic due to increased oxygen depletion by bacterial decomposition of planktonic blooms, the breakdown products become reduced rather than oxidised, and many of these products (e.g. hydrogen sulphide) produce offensive odours and tastes.

Excessive nutrient levels in aquatic systems can also cause two other kinds of ecological problems. Primarily, they may lead to extensive growth of aquatic weeds. Excessive growth of these weeds can impair fishing, bathing, fish spawning, shell fish production, and even navigation. Secondarily, nitrates can be converted in the human digestive tract by certain bacteria to nitrites, and the same transition may occur in opened cans of food even if they are subsequently refrigerated. Nitrites react with haemoglobin, forming methaemoglobin which will not take up oxygen. Laboured breathing and occasional suffocation result most severely in human infants. Nitrites may also react with creatinine (present in the vertebrate muscles) to form nitrosarcosine which can be carcinogenic.

EFFECTS OF ADVANCED EUTROPHICATION

1. Lush algae create problems of water colour, taste and odour, resulting in increased costs of water treatments.
2. The water is less attractive for boating, swimming and fishing.
3. The more desirable fish may be eliminated.
4. Irrigation canals may become clogged.

REMEDIES

The remedial measures to combat eutrophication include:

1. Removal of nutrients from waste waters, a very costly procedure.
2. Bypassing of lakes and diverting of waste waters to streams below the lake by pipeline.
3. Removal of excessive weeds and debris, and dredging of lake sediments.
4. Application of chemicals to destroy algal growth, copper sulphate and chlorine being commonly used for this purpose.

Review Questions

1. Define eutrophication and throw some light on its biological effects.
2. List out some remedies to combat eutrophication.

28

Waterborne Diseases

Water is contaminated by various sources, and it is evident that the microbes present in water are responsible for the outbreak of various diseases.

The most common waterborne diseases and their causative microbial agents are listed in Table 28.1

Several other categories of the disease other than those listed in Table 28.1 include:

1. Water-washed diseases are those resulting from inadequate personal hygiene because of scarcity or inadequate water supply, e.g. Typhus.

2. Water-based diseases are those arising from parasites which use an intermediate host that lives in or near water bodies.

3. Water-related diseases are borne by insect vectors which have habitats in or near water, e.g. Malaria

4. Water-dispersed diseases spread infections whose agents proliferate in freshwater and enter the human body, e.g. Legionellosis.

Table 28. 1   *Some important waterborne diseases*

Diseases and transmission	Microbial agent	Sources of agent in water supply	General symptoms
Amoebiasis (hand to mouth)	Protozoan (*Entamoeba histolytica* cyst like appearance).	Sewage, non treated drinking water, flies in water supply	Abdominal discomfort, fatigue, weight loss, gas pains, fever, abdominal pain, diarrhoea.
Campylobacteriosis (oral-faecal)	*Campylobacter jejuni*	Untreated water, sewage, living poor hygiene, crowded conditions with inadequate sewage facilities	Watery diarrhoea, vomiting, occasional muscle cramps
Cholera (oral-fecal)	*Vibrio cholerae*	Untreated water, sewage, living poor hygiene, crowded conditions with inadequate sewage facilities	Diarrhoea, abdominal discomfort
Cryptosporidiosis (oral)	Protozoan*Cryptosporidium parvum*	Collects on water filters and be membranes that cannot disinfected, animal manure, seasonal run-off of water	Flu like symptoms, watery diarrhea, loss of appetite, substantial loss of weight, bloating, increased gas

Table 28. 1   *Some important waterborne diseases (continued)*

Diseases and transmission	Microbial agent	Sources of agent in water supply	General symptoms
Cyclosporiosis	Protozoan *Cyclospora cayetanensis*	Sewage, non-treated drinking water aches, low-grade fever	Cramps, nausea, vomiting, muscle and fatigue.
Giardiasis (oral-fecal) and (hand to mouth)	Protozoan *Giardia lamblia* – most common intestinal parasite	Untreated water, poor disinfection, pipebreaks, leaks, ground water contamination, campgrounds where humans and wildlife use same source of water. Beavers and muskrats act as a reservoir for *Giardia*.	Diarrhoea, abdominal discomfort, bloating gas and gas pains.
Hepatitis A (oral fecal)	Virus – Hepatitis A	Raw sewage, untreated drinking water, poor hygiene, ingestion discomfort, jaundice, of shell fish from sewage flooded beds.	Fever, chills, abdominal urine dark
Salmonellosis (oral transmission)	Bacterium – *Salmonella* sp.	Contaminated water, shell fish, turtles, fish.	Gastroenteritis, fever and rapid blood-poisoning.

Table 28. 1 *Some important waterborne diseases (continued)*

Diseases and transmission	Microbial agent	Sources of agent in water supply	General symptoms
Shigellosis (oral–faecal)	Bacterium – *Shigella* sp.	Sludge, untreated waste water, ground water contamination, poorly disinfected drinking water	Fever, diarrhoea bloody stools
Schistosomiasis (immersion)	Schistosome	Contaminated fresh water with certain types of snails that carry schistosomes	Rash or itchy skin, fever, chills cough and muscle aches
Typhoid fever (oral–faecal)	Bacterium – *Salmonella typhi*	Raw sewage (carried and excreted in faeces by humans), water supplies with surface water source	Fever, headache, constipation, appetite loss, nausea diarrhoea vomiting and abdominal rash
Viral gastroenteritis (oral-faecal)	Viruses (includes Norwalk and rotavirus family)	Sewage, contaminated water, inadequately disinfected drinking water (mostly surface water sources)	Repeated vomiting and diarrhoea over 24 hour period. Gastrointestinal discomfort, headache, fever

Review Questions

1. What are the different categories of waterborne diseases?
2. Brief out the important waterborne diseases.

29

Purification of Water

The purpose of water treatment is to convert raw water into drinking water suitable for domestic use. Most important is the removal of pathogenic organisms and toxic substances like heavy metals which cause serious health problems. Other substances to be removed include suspended matter causing turbidity, iron and manganese compounds imparting a bitter taste and excessive carbon dioxide corroding concrete and metal parts.

Impure water may be purified by either of the following methods:

1. Natural
    (a) Storage
    (b) Oxidation and settlement
2. Artificial
    (a) Physical
        (i)   Distillation
        (ii) Boiling
    (b) Chemical
        (i)   Precipitation
        (ii) Disinfection or sterilisation
    (c) Filtration
        (i)   Slow sand filtration
        (ii) Rapid mechanical filtration
        (iii) Domestic filtration

Water purification may be considered under two headings:

PURIFICATION OF WATER ON A LARGE SCALE

Water on large scale, such as an urban water supply is purified in three main stages:

Storage

Water is drawn out from the source and impounded in natural or artificial reservoirs. Storage provides a reserve of water from which further pollution is excluded. As a result of storage, a very considerable amount of purification takes place. This is natural purification and we may look at it from three points of view:

a. *Physical* By mere storage, the quality of water improves. About 90% of the suspended solids settle down in 24 hours by gravity. The water becomes clearer. This allows penetration of light, and reduces the work of the filters.

b. *Chemical* Certain chemical changes also take place during storage. The aerobic bacteria oxidise the organic matter present in the water with the aid of dissolved oxygen. As a result, the content of free ammonia is reduced and a rise in nitrate occurs.

c. *Biological* A tremendous drop takes place in bacterial count during storage. The pathogenic organisms gradually die out. It is found that when river water is stored the total bacterial count drops by as much as 90% in the first 5-7 days. This is one of the greatest benefits of storage. The optimum period of storage of river water is considered to be about 10-14 days. If the water is stored for long periods, there is likelihood of development of vegetable growths such as algae which impart a bad smell and colour to water.

Filtration

Filtration is the second stage in the purification of water and quite an important stage because 98 to 99% of the bacteria are removed by filtration apart from other impurities.

The process of filtration may be of two types: (a) biological or slow sand filter and (b) mechanical or rapid sand filter.

Slow sand or biological filters   Slow sand filters for water treatment are in use since the 19th century. The elements of slow sand filter are depicted in Fig. 29.1 and include supernatant (raw) water, a bed of graded sand, an under-drainage system and a system of filter control valves.

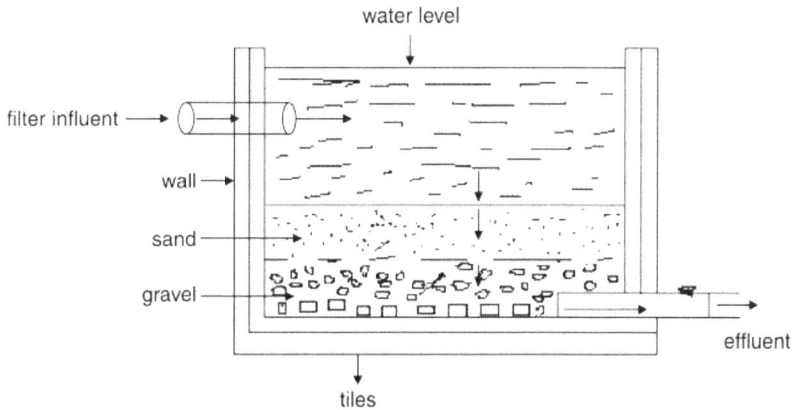

Figure 29.1   *Elements of slow sand filter*

1. *Supernatant water (raw water)*   The supernatant water above the sand bed, whose depth varies from 1–1.5 m serves two important purposes. It provides a constant head of water so as to overcome the resistance of the filter bed and thereby promote the downward flow of water through the sand bed. Secondly, it provides waiting period of some hours (3–12 hours), depending upon the filtration velocity) for the raw water to undergo partial purification by sedimentation, oxidation and particle agglomeration. The level of supernatant water is always kept constant.

2. *Sand bed/filter bed*   The most important part of the filter is the sand bed. The thickness of the sand bed is about 1.2 m. The sand grains are carefully chosen so that they are preferably rounded and have an 'effective diameter' (0.15–0.35 mm). The sand should be clean and free from clay and organic matter. The sand bed is supported by a layer of graded gravel which also prevents the fine grains being carried into the drainage pipes.

The sand bed presents a vast surface area. Water percolates through the sand bed very slowly (taking 2 hours or more) and as it does so, it is subjected to a number of purification processes¾mechanical straining, sedimentation, adsorption, oxidation and bacterial action, all playing their part.

The advantages of a slow sand filter are:

- Simple to construct and operate
- The cost of construction is cheaper than that of rapid sand filters
- The physical, chemical and bacteriological quality of filtered water is very high.
- It has been shown to reduce total bacterial counts by 99.9–99.99% and *E.coli* by 99 to 99.9%

Rapid sand or mechanical filters   It is popular in industrialised countries. Rapid sand filters are of two types, the gravity type and the pressure type. Both the types are in use.

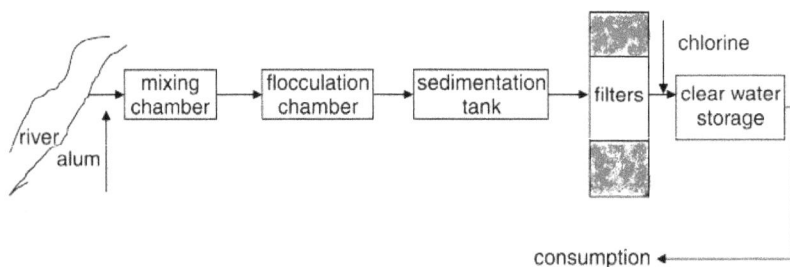

Figure 29.2   *Schematic representation of rapid sand filter plant*

The following steps are involved in the purification of water by rapid sand filters (Fig. 29.2).

1. *Coagulation*   The raw water is first treated with a chemical coagulant such as alum, the dose of which varies from 5–40 mg/mole/litre depending upon the turbidity and colour, temperature and the pH value of water.

2. *Rapid mixing*   The treated water is then subjected to violent agitation in a "mixing chamber" for a few minutes. This allows a quick and thorough dissemination of alum throughout the bulk of the water, which is very necessary.

3. *Flocculation*   The next phase involves a slow and gentle stirring of the treated water in a 'flocculation chamber' for about 30 minutes. The mechanical type of flocculator is the most widely used. It consists of a number of paddles which rotate at 2–4 rpm. The paddles rotate with the help of motors. This slow and gentle stirring results in the

formation of a thick, copious white flocculant precipitate of aluminium hydroxide. The thicker the precipitate or flock diameter, the greater the settling velocity.

4. *Sedimentation* The coagulated water is now led into sedimentation tanks where it is detained for periods varying from 2–6 hours when the flocculant precipitate together with impurities and bacteria settle down in the tank. At least 95% of the flocculant precipitate needs to be removed before the water is admitted in the rapid sand filters. The precipitate or sludge which settles at the bottom is removed from time to time without disturbing the operation of the tank. For proper maintenance, the tanks should be cleaned regularly; otherwise they may become a breeding ground for molluscs and sponges.

5. *Filtration*: The partly cleared water is now subjected to rapid sand filtration.

The advantages of a rapid sand filter over the slow sand filter are:

■ Rapid sand filter can deal with raw water directly. No preliminary storage is needed.

■ The filter beds occupy less space.

■ Filtration is rapid, 40–50 times that of a slow sand filter.

■ The washing of the filter is easy.

■ There is more flexibility in operation.

Chlorination

Chlorination is one of the greatest advances in water purification. It is a supplement, not a substitute to sand filtration. Chlorine kills pathogenic bacteria, but is has no effect on spores of certain viruses (e.g. polio, hepatitis) except when used in high doses. It oxidises iron, manganese and hydrogen sulphide; it destroys some taste and odour-producing constituents; it controls algae and slime organisms and aids coagulation.

Action of Chlorine When chlorine is added to water, there is formation of hydrochloric and hypochlorous acids. The hydrochloric acid is neutralised by the alkalinity of the water. The hypochlorous acid ionises to form hydrogen ions and hypochlorite ions as follows:

$$H_2O + Cl_2 \Longleftrightarrow HCl + HOCl$$

$$HOCl \rightarrow H^+ + OCl^-$$

The disinfecting action of chlorine is mainly due to the hypochlorous acid, and to a small extent due to the hypochlorite ions. The hypochlorous acid is the most effective form of chlorine for water disinfection. It is more effective (70–80 times) than the hypochlorite ion. Chlorine acts best as a disinfectant when the pH of water is around 7 because of the predominance of hypochlorous acid. When the pH value exceeds 8.5 it is unreliable as a disinfectant because about 90% of the hypochlorous acid gets ionised to hypochlorite ions.

Principles of chlorination   The mere addition of chlorine to water is not chlorination. There are certain rules which should be obeyed in order to ensure proper chlorination:

1.   The water to be chlorinated should be clear and free from turbidity. Turbidity impedes efficient chlorination

2.   The chlorine demand of the water should be estimated. The chlorine demand of water is the difference between the amount of chlorine added to the water, and the amount of residual chlorine remaining at the end of a specific period of contact (usually 60 min), at a given temperature and pH of the water. In other words, it is the amount of chlorine that is needed to destroy bacteria, and to oxidise all the organic matter and ammoniacal substances present in the water. The point at which the chlorine demand of the water is met is called the "break point". If further chlorine is added beyond the break point, free chlorine (HOCl and OCl) begin to appear in the water.

3.   The presence of free residual chlorine for a contact period of at least one hour is essential to kill bacteria and viruses. It should be noted however, that chlorine has no effect on spores, protozoan cysts and helminthic ova, except in higher doses.

4.   The minimum recommended concentration of free chlorine is 0.5 mg/l for one hour. The free residual chlorine provides a margin of safety against subsequent microbial contamination, such as may occur during storage and distribution. The sum of the chlorine demand of the specific water plus the free residual chlorine of 0.5 mg/l constitutes the correct dose of chlorine to be applied.

Method of chlorination   For disinfecting large bodies of water, chlorine is applied either as chlorine gas, chloramines or perchloron.

Chlorine gas is the first choice, because it is cheap, quick in action, efficient and easy to apply. Since chlorine gas is an irritant to the eyes and is poisonous, a special equipment known as 'chlorinating equipment' is required to apply chlorine gas to water supplies.

Chloramines are loose compounds of chlorine and ammonia. They have less tendency to produce chlorinous tastes and give a more persistent type of residual chlorine. The greatest drawback of chloramines is that they have a slower action than chlorine and therefore they are not being used much in water treatment.

Perchloron or high test hypochlorite (HTH) is a calcium compound which carries 60–70% of available chlorine. Solutions prepared from HTH are also used for water disinfection.

Break point chlorination   It is the name given to a method of chlorination where chlorine is added until the organic matter present in water is completely oxidised, and there remains a small quantity of "free" chlorine. The "point" at which free chlorine begins to appear is known as the break point. If further chlorine is added beyond the break point, it will result only in an increase of free chlorine. The 'free' residual chlorine may be estimated by the Orthotoluedine-Arsenite (OTA) test. Break point chlorination or free residual chlorination is the only reliable method of chlorination.

Superchlorination   This method which is followed by dechlorination involves the addition of large doses of chlorine after disinfection. This method is applicable to heavily polluted waters whose quality fluctuates greatly.

Other Agents Used in Purification of Water

While chlorine continues to be the most commonly used sterilising agent because of its germicidal properties and the comparatively low cost and ease of application, its pre-eminence in water disinfection is being seriously challenged because of the discovery that chlorination of water can lead to the formation of many 'halogenated compounds' some of which are either known or suspected carcinogens. As a result, many chlorine alternatives are receiving renewed interest. These include bromine, bromine-chloride, iodine and chlorine dioxide. As complimentary agents for chlorine in water disinfection, ozone is showing the greatest promise, and ultra-violet irradiation limited usefulness.

Ozonation   Ozone is a relatively unstable gas. It is a powerful oxidising agent. It eliminates undesirable odour, taste and colour, and removes all chlorine demand from the water. Most importantly, ozone has a strong virucidal effect. It inactivates viruses in a matter of seconds, whereas minutes are required to inactivate them with either chlorine or iodine. The drawback of ozone is that after it has done its job, it decomposes and disappears. There is no residual germicidal effect. The current thinking is that ozone should be used in the pre-treatment of water to destroy not only viruses and bacteria but also organic compounds that are precursors for undesirable chloro-organic compounds that form when chlorine is added. Thus ozonation is usually employed in combination with chlorination. The ozone required for potable water treatment varies between 0.2 and 1.5 mg/l.

Ultraviolet irradiation The germicidal property of UV rays has been recognised for many years. UV irradiation is effective against most microorganisms known to contaminate water supplies including viruses. The disadvantages are: (a) the process is generally more expensive than chemical treatment and (b) there is no residual germicidal effect. A further disadvantage is that colour and turbidity reduce the disinfection potential of UV rays. UV irradiation is not presently used on a large scale for water disinfection

PURIFICATION OF WATER ON A SMALL SCALE

Three methods are generally available for purifying water on an individual or domestic scale. These methods can be used singly or in combination.

Boiling

Boiling is a satisfactory method of purifying water for household purposes. To be effective, the water must be brought to a rolling boil for 5–10 min. It kills all bacteria, spores, cysts and ova and yields sterilised water. Boiling also removes temporary hardness by driving off carbon dioxide and precipitating the calcium carbonate. But it offers no 'residual protection' against subsequent microbial contamination. Water should be boiled preferably in the same container in which it is to be stored to avoid contamination during storage.

Chemical Disinfection

Bleaching powder Bleaching powder or chlorinated lime is a white amorphous powder with a pungent smell of chlorine. When freshly made, it contains about 33% of 'available chlorine'. It is however an unstable compound. On exposure to air, light and moisture, it rapidly loses its chlorine content. But when mixed with excess of lime, it retains its strength; this is called 'stabilised bleach'. Bleaching powder should be stored in a dark, cool, dry place in a closed container that is resistant to corrosion.

Chlorine solution   Chlorine solution may be prepared form bleaching powder. If 4 kg of bleaching powder with 25% available chlorine is mixed with 20 litres of water, it will give a 5% solution of chlorine. The solution should be kept in a dark, cool place in a closed container.

High test hypochlorite   HTH or perchloron is a calcium compound which contains 60–70% available chlorine. It is more stable than bleaching powder and deteriorates much less on storage. Solutions prepared from HTH are also used for water disinfection.

Chlorine tablets   Under various trade names (viz. halazone tablets) are available in the market. They are quite good for disinfecting small quantities of water, but they are costly.

Iodine   Iodine may be used for emergency disinfection of water. Two drops of 2% ethanol solution of iodine will suffice for one litre of clear water. A contact time of 20–30 minutes is needed for effective disinfection. Iodine does not react with ammonia or organic compounds to any appreciable extent hence it remains in its active molecular form, over a wide range of pH values and water conditions and persist longer than either chlorine or bromine. But its use invokes high costs.

Potassium permanganate   Once widely used, it is no longer recommended for water disinfection. Although a powerful oxidising agent, it is not a satisfactory agent for disinfecting water. It may kill cholera vibrios but is of little use against other disease organisms. It also alters the colour, smell and taste of water.

Filtration

Water can be purified on a small scale by filtering through ceramic filters such as Pasteur Chamberland filter, Berkefeld filter and Katadyn filter. The essential part of a filter is the 'candle' which is made of porcelain in the Chamberland type, and of kieselgurh or infusorial earth in the Berkefeld filter. In the Katadyn filter, the surface of the filter is coated with a silver catalyst so that bacteria coming in contact with the surface are killed by the 'oligodynamic' action of the silver ions, which are liberated into the water. Filter candles of the fine type usually remove bacteria found in drinking water, but not the filter-passing viruses. Filter candles are liable to be logged with impurities and bacteria. They should be cleaned by scrubbing with a hard brush under running water and boiled at least once a week.

Review Questions

1. What is the purpose of water treatment?

2. List out the various methods of water purification?

3. Describe in detail the purification of water on a large scale.

4. Brief out on the process of chlorination.

5. What are the principles of chlorination?

6. What is break point chlorination?

7. What is superchlorination?

8. Describe the agents other than chlorine used in purification of water.

9. Describe in detail the purification of water on a small scale.

10. What are the chemical disinfectants used to purify water?

11. Write a short note on filtration as an aid to water purification.

30

Recycling of Water

Water recycling is reusing treated wastewater for beneficial purposes such as agricultural and landscape irrigation, industrial processes, toilet flushing and replenishing a ground water basin (referred to as ground water recharge). Water is sometimes recycled and reused onsite as, for example, when an industrial facility recycles water used for cooling processes. A common type of recycled water is that which has been reclaimed from municipal wastewater or sewage. The term water recycling is generally used synonymously with water reclamation and water reuse.

NEED FOR WATER RECYCLING

Environment protection is universally acknowledged as one of the major issues facing the world today. The problem of water pollution due to its increased use in houses and industries in a country like India has reached the levels which renders powerless, the self purifying powers of nature.

Recycled water is waste water that has been treated to a level suitable for irrigation, industrial processing and other non-drinking uses.

Through the natural water cycle, the earth has recycled and reused water for millions of years. Water recycling, though, generally refers to projects that use technology to speed up these natural processes. Water recycling is often characterised as 'unplanned' or 'planned'. A common example of unplanned water recycling occurs when cities draw their water supplies from rivers that receive waste water discharges upstream from these cities. Water from these rivers has been reused, treated and piped into the water supply a number of times before the last downstream use withdraws the water. Planned projects are those that are developed with the goal of beneficially reusing a recycled water supply.

There are two grades of recycled water produced normally: one is used for industry and the other for irrigation. Recycled water is safe for human contact, but it is not intended for drinking. Recycled water has been used extensively throughout the United States¾including for food crops, over groundwater aquifers and in recreational lakes¾for the past 40 years with no negative health impacts.

Nature reuses and recycles its resources without creating toxic dumps or polluted waterways. The simplest form of water recycling is shown in Fig. 30.1. By imitating nature we can learn to recycle our wastewater relatively inexpensively, with minimal distribution costs, and the bonus of reclaimed resources ready to feed our fields or fish, or flush our toilets and water our lawns.

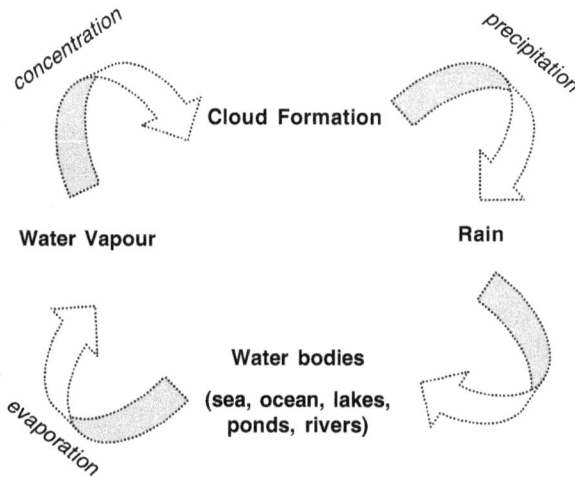

Figure 30.1 *Simplest Form of Water Recycling*

Throughout the world, alternative systems designed to mimic nature are operating to address the problems of conventional wastewater treatment facilities by making use of man-made ponds, constructed wetlands and designed soil filters to transform wastewater. At the Triangle School Wastewater Treatment Facility in Chatham County, North Carolina, water is being recycled for toilet flushing, thus helping the fishes to feed. Thus the wastewater is treated in a greenhouse, where it is used to produce plants and fish in an integrated cultivation system. In the case of less wealthy countries with serious health hazards resulting from insufficient sewage treatment facilities, affordable effective treatment alternatives incorporating large ponds and marshes are dramatically altering the quality of life. In Lima, Peru, water is treated by a series of ponds teaming with algae and organisms and fueled by sunlight. After 20 days, water is safe for reuse and feeds the fish ponds where phytoplankton gobble up remaining nutrients. Fish are harvested for human consumption and the sludge from the treatment ponds is used as a fertiliser on agricultural fields.

BENEFITS OF RECYCLED WATER

Recycled water can satisfy most water demands, as long as it is adequately treated to ensure water quality appropriate for the use. In uses where there is a greater chance of human exposure to the water, more treatment is required.

In addition to providing a dependable, locally controlled water supply, water recycling provides tremendous environmental benefits. By providing an additional source of water, water recycling can help us find ways to decrease the diversion of water from sensitive ecosystems. Plants, wildlife, and fish depend on sufficient water flows to their habitats to live and reproduce. The lack of adequate flow, as a result of diversion for agricultural, urban and industrial purposes, can cause deterioration of water quality and ecosystem health. Water users can supplement their demands by using recycled water, which can free considerable amounts of water for the environment and increase flows to vital ecosystems. Other benefits include decreasing wastewater discharges and reducing and preventing pollution. Recycled water can also be used to create or enhance wetland habitats.

REUSE OF DIFFERENT FORMS OF TREATED OR UNTREATED WATER

- Industrial use: Thermal power plant or wherever water is used for cooling
- Municipal use or domestic use
- Irrigation
- Agricultural use
- Potable-desalination of sea water
- Rain water harvesting.

Water recycling proves to be expensive hence it is practised only when:

- there is a large scale use of water which is inferior to the quality of available water supply.

- when the used water requires minimum treatment since it is not used as a raw material (thermal plants where it is used for cooling purpose only), sterilising, etc. (in food industries)
- processing plants where water can be used with minimum treatment like cleaning vegetables and washing vehicles.

The following waters can be recycled:

- Sewage waters
- Cooling tower waters
- Metal industry effluent recycling
- Engineering industries effluent recycling
- Textile industries effluent recycling
- Bottle wash effluent recycling
- Common effluent treatment plant discharge recycling

Waste waters are chiefly recycled using the methodologies of membrane separation technology. Microfiltration and ultrafiltration is a membrane fractionation process used to separate and concentrate micromolecules and colloids from water. These membrane separation systems are of spiral wound polyamide construction.

Water recycling has proven to be effective and successful in creating a new and reliable water supply, while not compromising on public heath. Non-potable reuse is a widely accepted practice that will continue to grow. Advances in wastewater treatment technology and health studies of indirect potable reuse have led many to predict that planned indirect potable reuse will soon become more common.

Review Questions

1. What is water recycling?
2. What is the need for water recycling?
3. What are the benefits of water recycling?
4. Which type of waters can be recycled?

PART VI

POPULATION INTERACTIONS

Microbial Interactions

Microbe-Microbe Interactions

Plant-Microbe Interactions

Animal-Microbe Interactions

31

Microbial Interactions

Organisms do not exist alone in nature but in a matrix of other organisms of many species. Thus interaction is the rule of nature. When one or more types of organisms reciprocate each others effect, it is called *interaction*. The evidence for such interaction is direct. Population of one species is different in the absence and in the presence of a second species. Interaction between two different biological population can be classified according to whether both population are unaffected by the interaction, one/both population benefit or one/both population are adversely affected.

CLASSIFICATION OF POPULATION INTERACTION

Name of interaction	Effect of interaction	
	Population A	Population B
Neutralism	0	0
Commensalism	0	+
Synergism (proto-cooperation)	+	+
Mutualism (symbiotic)	+	+
Competition	+	-
Amensalism	0/+	-
Parasitism	+	-
Predation	+	-

Now let us take a brief look at each of the interactions mentioned above.

Interactions are broadly classified based on their effects on each other as *positive interactions* and *negative interactions*. The positive interactions between biological populations enhance the ability of the interacting populations to survive within the community of a particular habitat. The mutualistic and synergistic interactions are some of the positive interactions between biological species. On the other hand, negative interactions between populations act as feedback mechanisms that limit population densities. In some cases, negative interactions may result in the elimination of a population that is not well-adapted for continued existence within the community of a given habitat. The negative feedback interactions limit population densities and provide a self-regulatory mechanism that benefits the overall population in the long run because it prevents overpopulation and destruction of the habitat's resources. Negative interactions also tend to preclude the invasion of an established community composed of autochthonous populations by allochthonous populations and thus act to maintain community stability.

Neutralism

It actually represents a lack of interaction between two populations. Though this kind of interaction is infrequent, yet it can occur between populations that are physically removed from each other. A lack of interaction is more likely at low population densities, when organisms are not likely to come into physical contact with each other, than when population densities are high, e.g. dormant resting stages of microbes are more likely to exhibit neutralism towards other microbial populations than are actively growing vegetative cells. Low rates of metabolic activity, which characterise the resting stages of microbes favour a lack of interaction.

Commensalism

In this kind of interaction among two populations, one population benefits and the other one is unaffected. It is a unidirectional relationship between populations. These kinds of interactions are quite common and they occur when the unaffected population modifies the habitat in such a way that a second population benefits, e.g. the removal of oxygen from a habitat, as a result of the metabolic activities of a population of facultative anaerobes, creates an environment that is favourable for the growth of obligately anaerobic populations.

Synergism

Synergism or proto-cooperation between two populations indicates that both populations benefit from the relationship but that the association is not obligatory. Both populations are capable of surviving independently, although they both gain advantage from the synergistic relationship. Such interactions are loose in that one member can be readily replaced by another. One form of synergism, syntrophism, occurs as a result of cross-feeding, in which the two populations supply each other's nutritional needs.

Mutualism

Otherwise known as symbiosis, mutualism is an obligatory interrelationship between two populations that benefits both of them. Mutualism is an extension of synergism, allowing populations to unite and establish essentially a single unit population that can occupy habitats unfavourable for the existence of either population alone. Mutualistic interactions may lead to the evolution of new organisms. It occurs when two populations are striving for the same resource. It usually focuses on a nutrient present in limited concentrations although it may also occur for other resources including light and space. As a result of this both populations achieve lower densities than would have been achieved by the individual populations in the absence of competition. Competitive interactions tend to bring about ecological separation of closely related populations, a fact known as the *competitive exclusion principle*. Competition prevents two populations from occupying the same ecological niche for the same nutrient resources.

Parasitism

In this kind of relationship, the parasite population is benefited and the host population is harmed. Parasitic relationships are characterised by a relatively long period of contact, and the parasite is smaller than the host. The host-parasite relationship is typically quite specific. Such host–parasite relationships that cause disease syndromes clearly exert a negative influence on the susceptible host and benefit the parasite.

Predation

It is another kind of relationship similar to parasitism but differing from it in the fact that here the predator is larger than the prey and that this kind of relationship is a short-term one and is not quite specific.

Review Questions

1. What is interaction?
2. Classify the various population interactions.
3. What are the effects of positive and negative interaction?
4. Describe the various population interaction.

32

Microbe–microbe Interactions

Microbes usually interact in a positive or negative manner. Neutralism is not frequent in the case of this kind of relationship. Among single population of microbes (microbes of the same species), microbes cooperate at low densities and compete at high cell densities. Beneficial interactions among microbes are facilitated by close physical proximity, e.g. biofilms and flocs.

POSITIVE INTERACTIONS WITHIN MICROBIAL POPULATION

Commonly, positive interactions dominate at low population densities and negative interactions dominate at high population densities. Hence there is an optimal population density for maximal growth rate. The various positive interactions within microbial populations are as follows:

Cooperation

Cooperation in a single colony of microorganisms is evidenced by extended lag period. For example, when a very small inoculum is used for culture, failure of growth results since there is an extended lag phase or a complete absence of growth. This is true for fastidious organisms and is a major obstacle to isolation procedures. Population of intermediate density are more successful than individual organisms for colonisation of natural habitats. Minimum infectious dose of pathogen or a single pathogen (a single cell) fails to cause disease.

How does cooperation occur? The semi-permeable membranes of microorganisms are imperfect and tend to leak low molecular weight metabolic products that are essential for biosynthesis and growth. Hence, within a population there are more concentrations of these extracellular metabolites that counteract the loss and facilitate re-absorption. But in a single cell or at very low population densities, loss exceeds replacement rates and prevents growth. A large population can adjust to an initially unfavourable culture medium but a very small population may be unable to do so. The basis of cooperation is colony formation. Motile bacteria that tend to move away often remain in colonies. Association of a population within a colony allows for more efficient utilisation of available resources and to seek new sources of nutrients (mass movement of colony). For example, in the slime mold, *Dictyostelium*, when the food resources become limited, the usually amoeboid colonies swarm together to form a central organism through the release of cyclic AMP. The cells unite to form a fruiting body which forms spores that disseminate and germinate to form the amoeboid colony. Thus during unfavourable conditions, the cells form a colony and tide over the condition. This is a fine example of microbial cooperation.

Commensalism

Mensa in Latin means table which further denotes 'one organism lives off the table scraps of another one'. Commensalism is a unilateral relationship and it results when the unaffected population modifies the habitat so that another population benefits. The modification of the environment by the facultative anaerobes for the obligate ones is an excellent example for this.

Commensalism is exhibited in a number of ways as listed below.

1. When there is a need for the production of growth factors commensalism plays an important role. For example, *Flavobacterium brevis* excretes cysteine which is taken up by *Legionella pneumophila.*

2. Transformation of insoluble compounds to soluble ones and further into gaseous components is observed in commensalism. For example, methane from sediment benefits methane oxidising population in the overlying water and hydrogen sulphide production in buried sediments which helps the photo-autotrophic sulphur bacteria on the sediment surface.

3. Conversion of organic molecules by one population which become substrates for other population is also commonly seen. For example, fungi release extracellular enzyme which decomposes lignin to form simple glucose moieties which benefit the other microorganisms.

4. Two organisms tend to remain together to satisfy their metabolic

needs. For example, *Mycobacterium vaccae*, growing on propane as a source of carbon and energy, will gratuitiously oxidise cyclohexane to cyclohexanone which is readily used up by other microorganisms like *Pseudomonas*. Here *Pseudomonas* benefit because they are unable to metabolise cyclohexane, whereas the *Mycobacterium* remains unaffected because it does not assimilate cyclohexanone.

5.   Removal or neutralisation of a toxic material is also observed. For example, oxidation of hydrogen sulphide by *Beggiatoa* detoxifies it, thus benefiting the H_2S sensitive aerobic organisms. Some tend to detoxify through immobilisation. For example, *Leptothrix* reduces manganese concentration thus permitting the growth of other microbes to which higher manganese concentration would be toxic.

Synergism

As already discussed above, proto-cooperation is a non-obligatory relationship between two organisms which satisfy each other's needs. It results in various kinds of actions: Syntrophism is a kind of synergism where the organisms satisfy their nutritional needs.

This is exhibited in a number of ways:

1.   Arginine is converted to putresciene by *E.coli* and *Enterococcus faecalis* both working together. *E.faecalis* can convert arginine into ornithine and *E.coli* further converts ornithine to putresciene, a task which cannot be performed by either of the organisms alone. Thus by remaining together both the organisms' nutritional needs are being fulfilled. In other words, both the organisms allow the completion of a metabolic pathway that otherwise would not be completed.

2.   Many allow supply of growth factors by one population for another. For example, *Nocardia* and *Pseudomonas* together can degrade cyclohexane wherein *Nocardia* acts on cyclohexane and supplies metabolic products to *Pseudomonas* and in turn the bacterium supplies biotin to *Nocardia*.

3.   Similar relationship is seen between *Lactobacillus arabinosus* and *Streptococcus faecalis* based on the mutual exchange of required growth factors. *L.arabinosus* requires phenylalanine for growth, which is produced by *S.faecalis* which in turn requires folic acid produced by *L.arabinosus*. In a minimal medium both populations can grow together, but neither population can grow alone.

4. In another relationship between *Chlorobium* (green alga) and *Desulfovibrio* (chemolithotrophic bacterium), *Chlorobium* fixes carbon dioxide through photosynthesis. It uses H_2S as the electron donor which it oxidises in the process to sulphur. This elemental sulphur is used by *Desulfovibrio* which reduces it to H_2S. It also makes use of the organic carbon supplied by the alga (for its growth) and supplies CO_2 which is taken up by the alga for photosynthesis.

5. Similar relationship can be seen among the bacterial population involved in nitrogen cycle. Heterotrophic pseudomonads are chemotactically attracted to organic excretions formed by heterocysts of *Anabaena spiroides*. They form dense aggregates around the heterocysts. Few bacteria are associated with photosynthetic cells of the filamentous alga. *Pseudomonas* oxidise the organic products released by the alga and thus stimulate nitrogenase activity (by lowering the oxygen concentration).

6. Some bacteria remain epiphytic on algal members taking up the preformed organic matter and oxygen (supplied by algae) and in turn supply carbon dioxide and vitamins for the algal members to flourish. Some synergistic relationships are based on the ability of second population to accelerate the growth rate of the first one. For example, the action of *Brevibacterium* and *Curtobacterium* which take up the excreted organic carbon given out due to metabolism of orcinol by *Pseudomonas*, helps in the acceleration of growth of *Pseudomonas* which would have otherwise declined in its growth due to accumulation of the metabolic products.

7. Synergistic relationships also allow microorganisms to produce enzymes that are not produced by either populations alone. For example, *Pseudomonas* strains (closely related species) produce lecithinase when grown together but not alone (as a single strain).

8. Degradation pathways of agricultural pesticides involve synergistic relationships. For example, *Arthrobater* and *Streptomyces* together degrade organophosphate insecticide Diazinon, a task which they cannot perform alone.

9. Another example is provided by *Penicillium piscarium* and *Geotrichum candidum* which remove toxic factors and produce useful substrates from insecticides. Both together detoxify the pesticide propanil. Penicillium cleaves propanil into propionic acid and 3,4–dichloroaniline (toxic) which is further detoxified by *Geotrichum*.

10. Archaeal population involved in methane production (methanogens) have interesting synergistic relationships with bacterial and other microbial population. The names of these bacterial genera indicate their syntrophic relationships with hydrogen consuming methanogenic archaea. *Syntrophomonas* species oxidise butyric acid and caproic acid to acetate and H_2. *Syntrophobacter* oxidises propionic acid to acetate, CO_2 and H_2. The acetate and H_2 produced by these bacteria are used by methanogenic archaea to produce methane. Thus the metabolism of the methanogens maintains very low concentrations of H_2 thus increasing the growth rates of *Syntrophomonas* and *Syntrophobacter* species.

Mutualism

It is an extended synergism which shows a highly specific and obligatory relationship. The mutualistic organisms lose their individuality and become a single organism. Some of the relationships are explained below:

Lichens   Lichens are organisms of symbiotic association in which a highly specialised fungal partner (mycobiont) inhabits in itself an algal partner (phycobiont). It is a thallus of undifferentiated tissues because of two symbionts. The fungal component mostly belongs to class Ascomycetes. Few Basidiomycetes members are also seen. It forms the main skeleton or framework of the lichen thallus. The algal component is either a member of green or blue-green algae (mostly single-celled green algae or filamentous blue-green algae). Some of the algal members which form symbiotic relationship with fungi are *Protococcus, Gloeocapsa, Nostoc, Rivularia,* etc.

Lichens are cosmopolitan in distribution, found in varied habitats from mountain tops to sea shores based on which they are of 3 different types: *Saxicoles* which grow on rocks, usually in cold areas. *Corticoles* which grow on barks of trees usually in the tropical and subtropical areas and *Terriclies* which grow in soil, in hot areas.

Lichens are varied in colours¾white, gray, yellow, black, green, etc. They do not thrive in or near polluted industrial areas. Lichens are perennial and slow growing and long lived. They can inhabit areas where normal plants cannot grow. For example, *Cladonia rangifera* grows in the Tundra region. They grow luxuriantly under moist conditions. Physiologically, lichens are a symbiotic unit of algae and fungi, algae supplying the photosynthates to the fungi and in return, fungi provide protection, water (absorbed from the moisture) and nitrogenous substrate, and anchors the lichen thallus firmly to the substratum.

On the basis of the fungal component, lichens are classified as *Ascolichens*, where Ascomycetes member is the fungal component (e.g. *Dermatocarpon*), and *Basidiolichens* where Basidiomycetes member is the fungal component (e.g. *Cora, Corella*).

On the basis of the distribution pattern of the partners, lichens are classified as *homoiomerous thallus,* where algae and fungi are uniformly distributed throughout the thallus (e.g. *Ephebe*), and *heteromerous thallus*, where algal cells are few in number and are distributed on the upper side of the lichen thallus in contrast to the fungal members that are dominant (e.g. *Parmelia*).

On the basis of the morphology, lichens are classified as *Crustose/Crustaceous*, where the lichen thallus is firmly adhered to the substratum and resembles crusts (e.g. *Verrucaria*); *Foliose*, where thallus has a flat leaf-like appearance which is slightly adhered to the substratum (e.g. *Gyrophora*) and *Fructicose*, where the lichen thallus is cylindrical or flat/ribbon-like resembling a little shrub and is very loosely adhered to the substratum (e.g. *Usnea, Cladonia*).

Endosymbionts of protozoa   Here algae and protozoa exhibit mutualism. *Paramaecium* hosts *Chorella* within its cytoplasm. Presence of *Chlorella* allows the protozoan to move into anaerobic habitats as long as there is sufficient light. Similarly the foraminiferans (protozoa) lodge Pyrrophycophycean members (red algae) within their cells. The algae impart red colour to the protozoans and each protozoan can contain about 50–100 algal cells. Fresh water protozoans usually lodge chlorophycean members within themselves and are known as zoochlorellae. Similar marine protozoan partners are Dinoflagellates which lodge chrysophycean members and are known as zooxanthellae. There are a few bacterial symbionts within protozoans. There is a unique relationship between *Paramaecium aurelia* and *Caedibacter*. *Paramaecium aurelia* exhibits two kinds of cells, one type called killer cells (contain endosymbionts, i.e. the *Caedibacter* whose cells contain an inclusion body called R body which is a toxin) and the other kinds are the sensitive cells. These killer cells compete with the sensitive strains (those devoid of the endosymbiont) for their survival. Endosymbiotic methanogens have been found in anaerobic ciliate protozoans living within the rumen, for example, *Methanobacterium, Methanocarpusculum* and *Methanoplanus*. They facilitate material exchange between the protozoan members and the bacteria.

Interaction of temperate phage and bacterial population   Genetic influence of phage particles incorporated within the genome of bacterial population (lysogenic phages) provide a mechanism of survival for the phage in a dormant stage for a long period. Bacteria harbouring lysogenic phages exhibit greater virulence. It also provides a mechanism for genetic exchange of bacterial DNA (transduction).

NEGATIVE INTERACTIONS WITHIN MICROBIAL POPULATION

The negative interactions with microbial populations include competition, amensalism, parasitism and predation.

Competition

Competition occurs when two populations use the same source of nutrition hence they achieve lower growth rates. Competition also brings about ecological separation of closely related population (competitive exclusion principle). Thus no two population can occupy the same niche. For example, in the case of *Paramaecium caudatum* and *P. aurelia* individually they maintain a constant population level but when the protozoans are placed together, it was seen that *P.aurelia* survived after 16 days. In another experiment two organisms (*P. caudatum* and *P. bursaria*) were artificially grown in conical flasks. Both were able to survive and reach a stable equilibrium even when grown together. It was evident that they competed for the same source but the two populations occupied different regions of the flasks. Thus in order to co-exist, the two population chose different habitats to allow the two species to grow together. Growth rates of competing population vary under different environmental conditions. In marine habitats, psychrophilic (cold loving) and mesophilic (that grow in moderate temperatures, i.e. 25–35°C) organisms occur together under same nutrient conditions. At low temperature, psychrophilic organisms grow and under moderately high temperatures, mesophilic organisms grow. At high substrate concentration, competition between a marine *Spirillum* and *E.coli* results in competitive exclusion of *Spirillum* whereas at low concentration of substrate, the reverse occurs and the *E.coli* is excluded. From this we can conclude that abiotic parametes such as temperature, pH, oxygen, etc. greatly influence the growth rates of microbial population.

Similarly, dominant microbial populations in sewage (having a high organic substrate content) are rapidly displaced in competition with the autochthonous microbial population of receiving streams and rivers, where the concentration of organic matter diminishes in the course of mineralisation and dilution. Under conditions of drought, the more tolerant species displace the more sensitive ones owing to the competitive ability of the former to survive. Competition is seen for multiple resources between two or more microbial species as in the case of *Microcystis aeruginosa* and the diatom *Asterionella formosa*. *Microcystis* utilises phosphorus as essential source but does not require silicon whereas *Asterionella* requires both silicon and phosphorus. Hence proportion of *Microcystis* decreases with an increase in silicon or phosphorus supply.

Amensalism

In amensalism, one microbial population growing on a substrate is inhibitory to the other population. This relationship is based on the production of certain microbicidal chemicals (allelopathic substances) or antibiotics. Amensalism leads to pre-emptive colonisation of a habitat, i.e. once an organism establishes itself within a habitat, it prevents other population from surviving in that habitat. For example, lactic acid bacteria prevent other microbial population from surviving in its substrate by producing large amounts of acids that prove detrimental to the other microbial population. Similarly, *E.coli* cannot survive in the rumen due to the production of volatile fatty acids produced by the already existing anaerobes. Fatty acids produced by microbes on the skin surface prevent the colonisation of these surfaces by other organisms like yeasts. Similarly, acids produced by microorganisms in the vaginal tract are responsible for preventing infection by *Candida albicans*. Some other examples of amensalism are: oxidation of sulphur by *Thiobacillus thiooxidans* which produces sulphuric acid which lowers aquatic pH thereby inhibiting many other microbes. Production of oxygen by algae may alter the habitat that can prove detrimental to obligate anaerobes.

Zymogenous populations grow under condition of high organic matter which permit the production of antibiotics. For example, *Cephalosporium graminerum* is a wheat pathogen that grows in dead wheat tissue and produces antifungal antibiotics to prevent attack by other fungi. Similarly *Trichophyton mentagrophytes* (a dermatophytic fungi) in the skin of New Zealand hedgehog produces penicillin (an antibiotic) on the skin and prevents the growth of the penicillin sensitive *Staphylococcus aureus* which is otherwise a common inhabitant of the skin. Bacteriocins are similar to antibiotics but their action is restricted to microorganisms very closely related to each other. They are peptides coded by plasmids and they are produced by one population to inhibit other microbial populations. For example, *Lactobacillus* species produce NISIN a bacteriocin that preserves food material like dairy products from spoilage by other bacteria.

Parasitism

It is a negative interaction wherein one population is adversely affected by the other. It is a long-term interaction and maintains a physical or metabolical contact. Parasites are the organisms which eventually kill one population (host) and they can be either attached outside the host (ectoparasite) or can be found within the host (endoparasite). Normally, parasitic interactions are very specific. A very good example for this kind of interaction is the virus population that parasitises a range of hosts like bacteria (bacteriophage), algae (phycophages), fungi (mycophages), actinomycetes (actinophages). A bacterium *Bdellovibrio* is ectoparasitic on several gram negative bacteria. Interaction occurs for one hour during which the host cell loses its shape and becomes spherical (bdelloplast). Other microbes cause lysis without direct contact. For example, Myxobacteria causes lysis of susceptible microorganisms with the help of exoenzymes. Some bacterial population produce chitinase which lyse the fungal cells. Protozoans are subjected to parasitism by fungi, bacteria and protozoa themselves. Algae are attacked by fungi (chitrids). Fungi attack fungi as seen in Agaricus being parasitised by *Trichoderma* species. Microorganisms that are themselves parasites may serve as a host for other parasites (hyperparasitism). For example, *Bdellovibrio* which is itself a parasite serves as a host for the phages. Thus parasitism provides a mechanism for population control and is density dependent, i.e. parasites can thrive only as long as the host survives.

Predation

This is another example for a negative interaction wherein one organism engulfs the other. The populations are known as predator (bigger) and the prey (smaller). Usually, the predator engulfs the prey and it is a short-term interaction in contrast to parasitism. Predator-prey relationship is cyclic and in each cycle, prey population increases.

The predator *Didinium nasutum* preys on *Paramaecium caudatum* whose population declines and becomes extinct. Once this happens, *Didinium* population also declines. If *Paramaecium* population is able to hide and escape predation, this population survives and increases in number.

Increased prey population

↓

Predator population follows
and overtakes the prey

↓

Decrease in the prey population

↓

Predator reaches low levels

↓

Prey population increases

↓

Predator also starts increasing

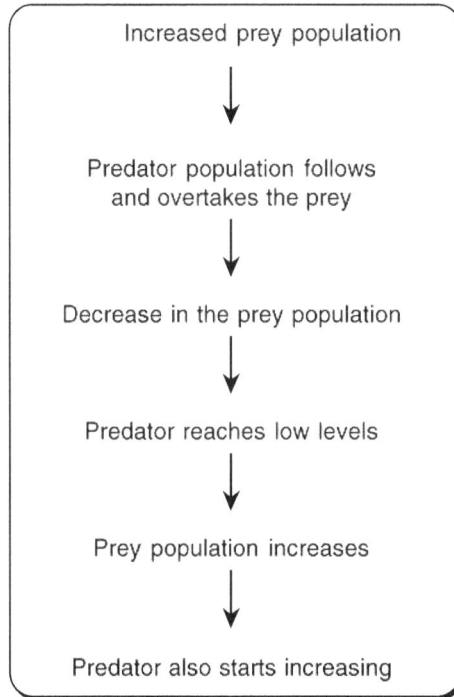

Review Questions

1. Define synergism and differentiate it from mutualism.
2. Define competitive exclusion principle.
3. Differentiate predation from parasitism.
4. How does cooperation occur?
5. What is syntrophism?
6. What are lichens? Describe in detail.
7. Discuss the Endosymbionts of protozoa.
8. Discuss the negative interaction within microbial population.
9. Define amensalism.

33

Plant-microbe Interactions

Positive and negative interactions take place not only between microbes but also between microbes and plants. The rhizosphere is a zone of predominantly commensal and mutualistic interactions between plants and microbes. Ecto- and endomycorrhizal fungi provide plants with mineral nutrients and water, receiving photosynthates in return. Under harsh conditions, this mutualistic association can be essential for plant survival. The associations of dinitrogen-fixing bacteria with certain plants provide essential combined nitrogen for crops and ecosystems. The aerial surfaces of plants provide habitats for largely commensal microbes. On the negative side, certain viruses, bacteria, and fungi cause plant diseases that can result in great economic losses and even severe food shortages.

Some of the positive interactions among plants and microbes are:

- Synergistic interactions which include rhizosphere, rhizoplane, phyllosphere and spermosphere.
- Mutualistic interactions which include root nodule interactions, leaf nodule interactions and mycorrhizal interactions.

Parasitism is the only negative interaction among the plant and the microbes.

SPERMOSPHERE

It is the volume of soil that surrounds a seed. It is an area of increased microbial activity around a germinating seed because of the nutrients leaked into the soil by the germinating seeds. In 1904, M. Duggali was the first to report about the bacterial flora carried by healthy seeds. While most of the microbes are harmless, some may be positively beneficial and some of them may be pathogenic.

Spermosphere organisms forming the normal flora around a germinating seed have some beneficial effects through biological products like growth hormones. Germinating seeds excrete certain chemicals which may influence the quality and quantity of microbes in the vicinity of the seed.

Spermosphere Effect

When a seed is sown in soil, certain interactions take place between the seed-borne microflora (due to the secretion of certain chemicals by the seed) and the soil-borne microflora which influence the quality of the spermosphere at that condition. When the seed is pre-treated with a fungicide or with any other biological agent, this influences such interactions to a great extent, as for example, the fungicide may totally alter the seed microflora (by inhibiting some fungal flora and increasing some other bacterial flora). This could also influence the nature of microflora that is about to colonise the root (rhizosphere) once the radical emerges out of the seed. Thus by manipulating the spermosphere, one changes the rhizosphere also.

When a seed carrying a natural or altered (by manipulating) load of microbes is sown, certain microbes are activated and others are suppressed. Usually, the microbes that are artificially loaded onto the seed are more dominant flora of the seed and this enables the scientists to beneficially alter the spermosphere of a particular seed, as in *Rhizobium*, *Azotobacter* and *Azospirillum* coated seeds. These organisms get established on the root surface of the germinating seed and benefit the plant (by fixing nitrogen directly into the roots).

Along with spermosphere microflora, the soil-borne flora may also get activated and compete with the former (for nutrition and space). The qualities of chemicals excreted by the germinating seed decides the final quality and quantity of the microflora around the seed. Usually microbes move from the spermosphere to the rhizosphere within three days. Various chemical treatments of the seed (organo-mercurial pesticides) definitely change the rhizosphere microflora of the seedling thus indicating the plant–root–microbes interactions in the soil through the seed.

When the seed is internally or externally infected by certain pathogenic microorganisms (smut spores), this definitely alters the quality and quantity of the spermosphere and rhizosphere microflora (again through competition). When such a suspected seed is pre-treated with plant protection chemicals, the competition is eliminated (since the pathogen gets killed) and hence the seed gets coated with harmless microbes.

When the seed is pre-treated with organic manure (cow dung) where the seed gets coated with saprophytes present in the manure there is competition between pathogens and non-pathogens (saprophytes present in the organic manure) and depending on the efficiency of one group, one is suppressed and gets eliminated. For example, a seed-borne pathogen of cotton *Xanthomonas campestris* pv *malvacearum* is controlled when the seed is pre-treated with cow dung slurry containing a lot of saprophytes.

To understand the effect better, we can take this example. When a seed infected with a pathogen is sown in unsterile and sterile soil, there is intense spermospheric effect (enough to suppress the pathogen) in the former whereas in the latter case, pathogen becomes highly virulent (since there is no competition by other organisms).

RHIZOSPHERE

Rhizosphere is the region where soil and roots of the plants make contact or the thin layer of soil adhering to a root system after shaking and removing the loose soil.

Rhizosheath is a modification of rhizosphere, characterised by a relatively thick soil cylinder that adheres to the plant roots. This is typical of some desert grasses.

Rhizoplane or root surface   When the roots are cleaned of all the soil particles adhering to it and then plated, microorganisms can be seen developing indicating that there are certain microbes intimately associated with the root surface. Some fungi inhabit the root surface in a mycelial state, e.g. *Cephalosporium, Trichoderma, Penicillium*. Specific bacteria also get embedded on the surface of the root with the help of mucilaginous external layer normally present in the actively growing root system.

Rhizosphere effect   The direct influence of plant roots on microbes and microbes on plant roots within the rhizosphere is known as the rhizosphere effect.

Effect of Plant Root on Microbial Populations

The structure of the plant root system contributes to the establishment of the rhizosphere microbial population. The interactions of plant roots and rhizosphere microorganisms are based largely on interactive modification of the soil environment by processes such as water uptake by the plant and release of organic chemicals by the roots. The influence of the plant root on the microflora is governed by

- root exudates
- physical and chemical factors in the soil

Effect of root exudates   This is the major factor that governs the microflora of the rhizosphere. The root exudates include:

- Simple sugars such as glucose and fructose
- Di, tri and oligo saccharides
- All common amino acids — alanine, serine, leucine, valine, glutamic and asparitic acids. Of these, glutamine and asparagine are produced in large amounts.
- Vitamins¾thiamine and biotin
- Nucleotides
- Flavones and auxins
- Stimulators/inhibitors of particular microbes

All these root exudates have an effect on the rhizosphere microflora. Some of the root exudates like the amino acids, promote the growth of microflora of the rhizosphere.

Some nitrogen fixers such as *Azospirillum*, *Azotobacter paspali* use the root exudates as the energy source for significant nitrogen fixation.

Thus there is a distinct selective influence of the root system over the microbes. For example, there is a preferential stimulation of gram negative non-spore forming rods in the root region.

Root exudates containing toxic substances such as glycosides and hydrocyanic acid may inhibit the growth of pathogens.

One of the attributes of root exudates is the possible role they play in neutralising the soil pH and altering the microclimate of the rhizosphere through liberation of water and CO_2. Such changes may influence infections of roots by pathogenic fungi.

Effect of plant growth on rhizosphere microflora   The rhizosphere microflora may undergo successional changes as the plant grows from seed germination to maturity. During plant development, a distinct rhizosphere succession results in rapidly growing, growth factor-requiring, opportunistic microbial population. These successional changes correspond to changes in the materials released by the plant root to the rhizosphere during plant maturation.

Initially, carbohydrate and mucilaginous exudates from plant roots stimulate the growth of microorganisms rapidly within the grooves on the root surface and within the mucilaginous sheath (rhizoplane). After the plant matures, autolysis of some of the root materials take place and simple sugars and amino acids are released into the soil. This further stimulates the growth of bacteria with high intrinsic growth rates, e.g. *Pseudomonas*. As a result of these effects, the rhizosphere microflora consists of higher proportion of gram negative rods and a lower proportion of gram positive rods, cocci and pleomorphic forms. A relatively higher proportion of motile, rapidly growing bacteria are also seen.

Alteration of rhizosphere microflora   This may be done by:

- *Soil amendments*   This refers to the artificial addition of fertilisers with nitrogen, phosphorous and potassium. This depends on the rhizosphere : soil (R:S) ratio and also on the nutritional content of the chemicals in the soil.

- *Foliar application of nutrients*   Translocation of photosynthates from leaves to roots is a well known phenomenon and this does not affect the microflora. So when foliar application of antibiotics, growth regulators, pesticides and inorganic nutrients is carried out, a small amount is being released as root exudates and this can either promote the growth of the present microflora or change the microflora to some extent.

- *Artificial inoculation*   This is done on seed or soil with preparation containing live microorganisms especially bacteria (*bacterisation*). This is beneficial, in that this provides an easier way for the establishment of the microbes to the rhizosphere. This is because as the seed is coated with the live microorganisms, as soon as the root evolves, the colonising of the root takes place and establishment of the other microbes is also made possible. Microbial seed inoculants generally used are *Azotobacter, Beijerinckia, Rhizobium* or phosphorous solubilising microorganisms.

Effect of Rhizosphere Microbial Population on Plants

The microorganisms have a marked influence on the growth of plants. The plant growth may be impaired due to the absence of appropriate rhizosphere microflora. The microbial population affect the plant growth in various ways:

Promotion of growth   This is brought about by the release of growth factors like auxins and gibberellins that promote plant growth. The organisms which release these growth factors include *Arthrobacter, Pseudomonas* and *Agrobacterium*. The production of indole acetic acid (IAA), a plant growth hormone by certain group of microorganisms increases the rate of seed germination and development of root hairs. This is seen in wheat seedlings.

Neutralisation of toxic substances   This is seen in the case of plants that grow in flooded sediments, e.g. rice plants and other partially submerged plants. In this case, there is production of hydrogen sulphide generated by the sulphate reduction pathway. This hydrogen sulphide is toxic to the plant roots, and this is neutralised by the bacteria *Beggiatoa*. This is a microaerophilic, catalase negative, sulphide oxidising filamentous bacterium. This acquires the oxygen and catalase enzyme from the rice plant and aids in the oxidation of toxic H_2S to harmless sulphur or sulphate, thus protecting the rice roots.

Allelopathic effect   Some substances being released by the microbes can have an antagonistic effect. This may allow plants to enter in amensalic relationship with other plants. Some substances or extracellular products of certain microorganisms lead to the growth of other kinds of microorganisms that can provide a better rhizosphere microflora. These extracellular products can also inhibit the growth of pathogens, thus protecting the plant roots from getting damaged.

Nutritional recycling   The nutrients in the soil are made available to the plants by mobilisation of the nutrients, by fixing it in soil in proper way. Sometimes, the nutrients are made unavailable by immobilisation. For example, microorganisms produce extracellular amino acids, vitamins, etc. using the nutrients and nitrogen fixation process. Thus they make nitrogen available to plants as nitrates or other inorganic forms, e.g. *Rhizobium* and *Azotobacter*. Similarly, sulphur oxidisers make sulphur available as sulphates, e.g. *Desulfovibrio*. Phosphorous is made available as phosphates by the production of acids by the microflora. Siderophore production is another important characteristic feature of rhizosphere microflora.

Siderophore production   Many microorganisms respond to a fall in the availability of iron in soil by producing extracellular low molecular weight iron transporting agents known as siderophores. These siderophores selectively complex with iron and supply the element to the living cell. They also act as growth factors or antibiotics. For example, *Pseudomonas fluorescence* (one strain produced a siderophore compound *pseudobactin*) inhibits the growth of a pathogen *Erwinia carotovora* by chelating iron from the vicinity of the pathogen and thus reducing the disease severity.

Thus microorganisms increase the recycling and solubilisation of mineral nutrients and making it available to plants. The abundant growth of microbial population in the rhizosphere can sometimes create a deficiency of required minerals for the plants, e.g. bacterial immobilisation of zinc and oxidation of manganese cause the plant diseases 'little leaf' of fruit trees and 'gray speck' of oats. Nitrogen is immobilised in the form of microbial protein and some may be lost to the atmosphere by denitrification.

PHYLLOSPHERE

The Dutch microbiologist Ruinen coined the term *phyllosphere* which is the interrelationship between plant foliage and the quality and quantity of microorganisms found on the surface.

The leaf surface is termed as *phylloplane*. The quality and quantity of the microorganisms on the leaf surface differs with age of the plant, leaf area, morphology, atmospheric factors (temperature, humidity, etc.). Growing seasons may also influence the phyllosphere microflora. It increases and reaches the maximum in autumn when the leaves severe. The position of the leaves also plays a role in determining the microflora.

Leaves at the lower levels harbour more microorganisms since they are sheltered and get more nutrients from the raindrops from upper levels.

Plant leaves are exposed to dust and air currents resulting in the establishment of a typical flora on their surface aided by cuticle, waxes and appendages (thorns, spikes) which help in the anchorage of microbes. The leaf diffusates/exudates promote/deter the growth of microbes on their surface. The principal nutritive factor in the leaf are amino acids, glucose, fructose and sucrose.

The dominant microorganisms in a forest vegetation are the nitrogen fixing bacteria such as *Beijerinckia* and *Azotobacter*. Other genera like *Pseudomonas, Erwinia, Sarcina* have been encountered in the phyllosphere.

Under damp conditions, some leaves may harbour cyanobacteria like *Anabaena, Calothrix, Nostoc* and *Tolypothrix* on their surfaces.

Some of the fungi and actinomycetes encountered are *Cladosporium, Alternaria, Cercospora, Helminthosporium, Mucor* and *Streptomyces* species.

Characteristic Features of Phyllosphere Microflora

■ Leaf surface microbes may perform an effective function in controlling the spread of airborne microbes inciting plant diseases.

- Presence of a fungal spore on the surface of leaves incite the formation of a chemical substance referred to as *phytoalexin* which are active in host defence mechanisms.

- Resistance to disease causing microbes has also been attributed to fungistatic compounds secreted by leaves such as malic acid from leaves of *Cicer arietinum*.

- The name *elicitor* has been commonly used to denote the compounds which induce the synthesis of phytoalexins. These are biotic elicitors such as polysaccharides from fungal cell walls, lipids, microbial enzymes and polypeptides.

- Abiotic elicitors are heavy metal salts, detergents, UV light, etc.

- Epiphytic microbes are known to synthesise indole acetic acid.

- Phyllosphere bacteria are often pigmented due to direct solar radiation.

- Pink-pigmented facultative methylotrophs are common in the phyllosphere.

- Bacteria can serve as ice nucleators, promoting frost damage to plants. Genetically modified *Pseudomonas syringaea* lacks a membrane protein that promotes nucleation. Inoculating crops with this organism can lower the temperature at which frost damage occurs.

- Azolla-Anabaena symbiosis is a N_2-fixing association where cyanobacteria live on leaf surface.

Any change in phyllosphere affects plant growth which in turn affects the physiological activity of root system. Such changes in the root result in an altered pH and spectrum of chemical exudation causing a change in rhizosphere microflora. Thus there is a link between phyllosphere microflora and rhizosphere microflora. There is a continuous diffusion of plant metabolites from the leaves which support the microbial growth and in turn these microbes protect the plant from pathogens.

There are reports on suppression of phyllosphere microflora due to environmental pollution caused especially by cement and fertiliser industries.

POSITIVE INTERACTIONS OF PLANTS AND MICROBES

Bacterial mutualistic interactions with plants have been seen in the leaf nodules. Fungal mutualistic interactions are pronounced in plants.

LEAF NODULE

Symbiotic association of certain bacterial endophytes with leaves of certain plants (usually belonging to the families Rubiaceae and Myrsinaceae) leads to the formation of nodule like structures on the leaves. The plants *Psychotria, Pavetta, Chomelia* have received considerable importance as leaf nodule producing plants.

Isolates of bacteria that have gained importance in formation of leaf nodules are *Mycobacterium rubiacearum, Mycoplana rubra, Flavobacterium* species, *Bacterium rubiacearum, Phyllobacterium rubiacearum* and *Klebsiella rubiacearum.*

There are not many advantages (for the plants) resulting from this sort of mutualistic interactions apart from the fact that the bacterial partners secrete phytohormones (cytokines) for the growth of the plants. The plants definitely provide shelter and photosynthates for the survival of their bacterial partners. Since they form a stable phyllospheric microflora with the plants, these bacterial partners may also prevent the entry of pathogenic spores from entering through the leaves and establishing themselves.

MYCORRHIZA

The term 'mycorrhiza' (literally, *fungus root*) was first used by A.B.Frank to characterise the association between higher plants and fungi. The symbiotic association between the roots of some plants and some fungi is called mycorrhizal association. The fungus is highly habitat limited and is usually found in the immediate vicinity of or within the roots.

Occurrence of Mycorrhiza

They occur on almost all terrestrial plants though not as specific as the nitrogen fixing symbiosis. Thus a plant may have several mycorrhizae that can form symbiosis with its roots. Extent of symbiosis depends on fertility. High soil fertility leads to low mycorrhizal infection and poor symbiosis and vice versa. Roots supply carbohydrates to the fungi which absorb nutrients form the soil and supply them to the crop.

Types of mycorrhizae

Mycorrhizae are of two kinds:

Ectomycorrhiza Ectomycorrhizal symbiosis is a mutually beneficial union between fungi and the roots of vascular and non-vascular plants. The host of an ecomycorrhizal fungi is usually a gymnosperm (pine). The typical ectomycorrhizal fungi are basidiomycetes (Agaricus), ascomycetes or phycomycetes members. Usuallly more common in the temperate regions, they are capable of growing apart from the host on media containing simple sugars and vitamins. Ectomycorrhizae have poor competitive saprophytic ability hence they have a tough time competing with other microbes in the soil.

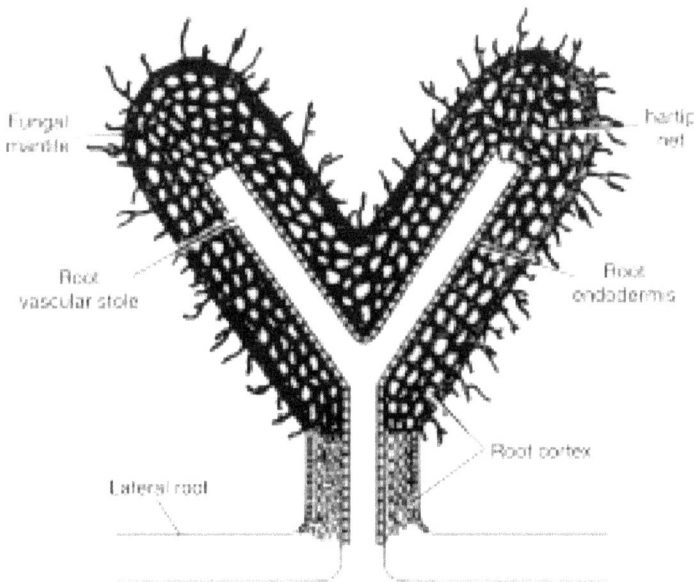

Figure 33.1 *Ectomycorrhizal association*

During the infection process, ectomycorrhizal fungi in soil are stimulated by the root metabolites to grow toward the root (Fig. 33.1). The hyphae aggregate around the root and penetrate between the root epidermis and the cortex. A structure called *Hartig net* is formed which is a fungal sheath surrounding the root in which fungal hyphae penetrate between the root cells. Eventually, the root gets surrounded by a fungal mantle. The fungal hyphae replace the fine lateral root hairs of the host root system thus modifying it structurally. Thus the host root system that is infected with ectomycorrhiza looks stunted and dichotomously branched. The fungal hyphae on the exterior of the roots usually serve as an extension of roots and store large amounts carbohydrates.

Endomycorrhiza In this case, the fungal hyphae penetrate the host root cells. They are quite common among the Ericaceae and Orchidaceae members of higher plants as well as fruit trees like citrus, coffee, rubber, etc.

VAM FUNGI

Vesicular-arbuscular mycorrhiza (VAM) fungi shown in Fig. 33.2 are geographically ubiquitos. They are commonly found in association with agricultural crops, shrubs, tropical tree species and some temperate trees. Their nutritional requirements are not specific. VAM associations are formed by non septate Zygomycetes and Phycomycetes fungi. Some examples are *Glomus, Gigaspora, Acaulospora, Entrophospora* and *Scutellospora* of which *Glomus* is the most common fungus. The fungi are obligate biotrophs and do not grow on synthetic media and hence are classified according to the morphological characteristics of the spores formed in the soil. VAM fungi produce large resting spores (0.2 mm). The spores can survive adverse conditions. The germ tube dies if it is unable to encounter and successfully penetrate the host root. The fungal hyphae traverses and ramifies in the root cortex. Branches from the intercellular hyphae enter cortical cells where further branching results in highly branched hyphal structures called arbuscles. They are short-lived and serve as the nutrient transfer

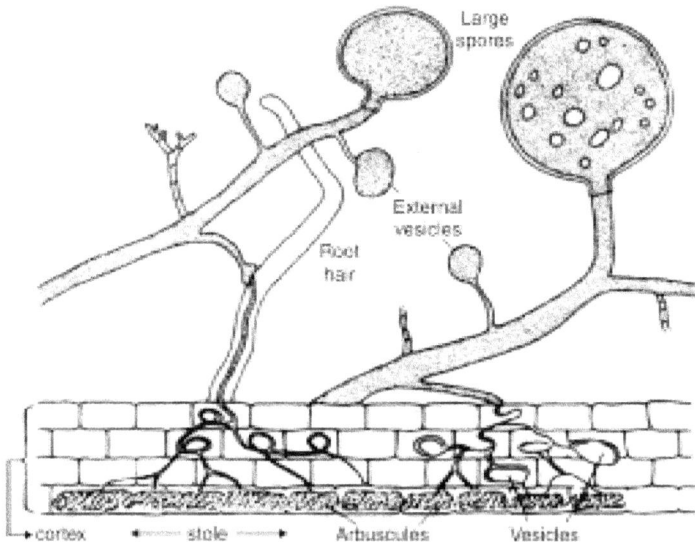

Figure 33.2 *VAM fungi*

mechanism between the fungus and the host. Phosphate transfer possibly occurs across living membranes of the host and the fungus via arbuscles. When the association is well established, hyphal swellings called vesicles form on the mycelium inside and outside the root. The vesicles are sac-like terminal swellings at the tip of hyphae and contain many lipid droplets and function as storage organs. External mycelium form very thick-walled chlamydospores.

These fungi are disseminated both actively and passively. Active dissemination is form one root to another by mycelial growth through soil and passive dissemination is through biotic agents like rodents, worms, insects and birds or abiotic agents like wind and water.

VAM fungi interact with other soil microbes like the free-living and symbiotic nitrogen fixers and phosphate solubilisers to improve their efficiency for the biochemical cycling of elements to the host plants.

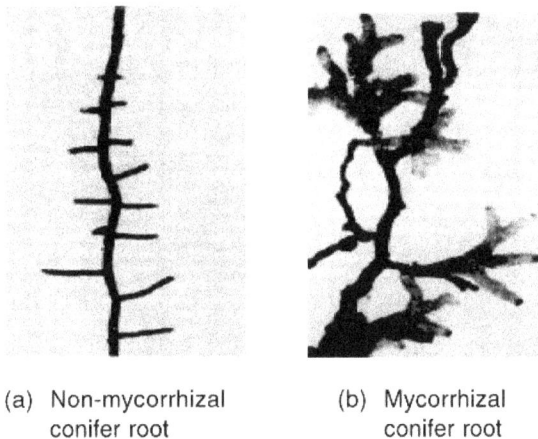

(a) Non-mycorrhizal   (b) Mycorrhizal
    conifer root   conifer root

Figure 33.3 *Forked short roots*

Figs. 33.3 (a, b) show forked short roots caused by an ectomycorrhizal fungus.

NEGATIVE INTERACTION BETWEEN PLANTS AND MICROBES

As already discussed, parasitism is the only negative interaction between the plants and microbes.

As parasites, the microbes, like bacteria, fungi, viruses and algae, cause infections in the host plant leading the development of disease and loss of commercial value in case the host plant is an agricultural crop.

To cause the disease, the parasite must accomplish two important things:

1. It must enter the host plant.
2. It must establish itself at the specific target site within the plant.

After accomplishing this, the parasite is able to overcome the plant defence mechanisms and causes the disease.

There are different portals of entry for each microbe. Thus bacteria and viruses enter the host tissue through the natural openings like stomata, lenticels, etc. Fungi have a separate mode of entry. Either the fungal spore enters directly through the natural openings and germinates within the host tissue, or the spore on falling on the plant surface, forms a special adaptation called *appresorium* which anchors the spore onto the substratum. From the appresorium arises a small projection called the *infection peg* which has a sharp tip through which it penetrates the host cell wall actively.

The pathogens can also enter through open wounds (scratches) on the plant surfaces.

Once the pathogen enters the host tissue, it tries to overcome the host defence mechanisms (production of phytoalexins on induction by the elicitors which are the pathogens themselves) and establishes relationship with the host. This interaction is of two kinds. One is *biotrophic interaction*. Here the pathogen enters into a harmonious relationship with the host with the pretext that it continues to obtain nutrition from the plant for a long time. Thus it does not kill the plant instantaneously. Another relationship is the *necrotrophic* wherein the pathogen on entering the host kills the host and obtains the nutrition from the dead tissues.

After the disease is established and even during the course of the disease establishment, a series of symptoms can be visualised on the host surface which is an indication of disease onset.

Some examples of parasites on plants are:

*Bacteria*¾*Erwinia carotovora* causing soft rot in carrots; *Xanthomonas campestris* pv *oryzae* causing bacterial blight in rice.

*Fungus*¾*Fusarium oxysporum* causing wilt disease in many plants (cotton, banana).

*Virus*¾Tobacco mosaic virus causing mosaic disease in tobacco.

Review Questions

1. What is rhizosphere?

2. What is spermosphere?

3. Describe the spermosphere effect.

4. Describe rhizosphere as an interaction between plants and microbes.

5. What is rhizosphere effect? Explain.

6. How can you alter the rhizosphere microflora.

7. Explain the effect of rhizosphere microflora on plants.

8. What is allelopathic effect?

9. Throw some light on siderpohore.

10. What is phyllosphere? Explain.

11. Describe leaf nodule as a positive interaction.

12. Define mycorrhiza.

13. Explain in detail mycorrhizal interactions and their beneficial effects.

14. Describe the negative interaction between plants and microbes.

34

Animal-microbe Interactions

Most interactions between microbes and animals are beneficial. The mutualistic relationships of microbial and animal population involve nutrient exchange and maintenance of a suitable habitat. These associations help the animals to digest difficult components of their diet particularly cellulose.

Intestinal symbionts may be commensals or benefit the animal through vitamin production and protection against pathogens. The endozoic algae of coral polyps and other invertebrates supply a major part of the animal's nutritional needs through their photosynthetic activity. Associations with chemoautotrophic bacteria in deep-sea thermal vent environment allow invertebrates to live on geothermal energy independent of photosynthetically produced organic carbon. In a less common mutualistic relationship, endosymbiotic bacteria produce light for some marine invertebrates and fish.

PREDATION OF MICROBES BY ANIMALS

Many invertebrates satisfy part of their food requirements by preying on microbes 10^5-10^7 times smaller in biomass than themselves. This is accomplished by two feeding strategies:

Grazing   This is a common feeding strategy of aquatic invertebrates like snails and sea urchins. They scrape and ingest the microbial crust from submerged surfaces where the microbial populations are able to reach high densities because of the physical absorption of dissolved nutrients on these surfaces. The size difference between the predator and prey becomes relatively unimportant in this feeding process (grazing) because the predator pursues masses of millions of microbes rather than individual prey.

Various members of the marine habitat, secrete slime trails which bacterial, fungal and algal populations colonise. These slime trails provide nutrition for the microbial population. They adhere to the slime which acts as mucous traps. The animals then retrace their tracks and graze on the microbial population that are entrapped in the mucous slime trails.

Filter feeding   Many sessile benthic (sediment) invertebrates that are permenantly attached underwater exhibit this mechanism to exploit suspended planktonic microbial prey. These animals remain stationary and filter the prey out of suspension. This process is energetically advantageous to the predators which maintain a flow of water using cilia through which the microbes are filtered. For example, sponges, bivalve molluscs and barnacles exhibit this kind of feeding.

The microbes associated with faecal pellets are an important source of food for many aquatic and terrestrial animals. Digestion of food during passage through the alimentary canal is usually incomplete leaving the cellulose portion undigested. The excreted faecal pellets are further decomposed by the remnants of intestinal microbes and additional microbes from the environment when the recalcitrant plant polymers are solubilised and converted to microbial biomass. Re-ingestion of the faecal pellets by the same animal or other animal population allows a more complete utilisation of the food resource. In addition, the microbes also supply critical vitamins than otherwise would be absent in the diet. In the terrestrial habitat, certain rodents (rabbits) are *coprophagous* (re-ingest their own faecal matter).

Most herbivorous animals are unable to digest the cellulosic parts of the plant materials they consume. They rely on the enzymatic capabilities of microbes to degrade this material and to produce substances that they can assimilate.

INTERACTION BETWEEN MICROBES AND THE GUT

Most warm-blooded animals contain extremely complex microbial communities within their gastrointestinal tracts. In the human intestine, the strict anaerobes belonging to the genera *Bacteroides*, *Fusobacterium*, *Bifidobacterium* and *Eubacterium* are found in large numbers. In some animals such as pigs, the gastrointestinal tract microflora contribute to the nutrition of the animal by fermenting carbohydrates. The other activities include amino acid degradation, supply of required vitamins by the microorganisms, e.g. synthesis of vitamin K. The presence of gut microflora and the pre-emptive colonisation, constitute an important barrier to attack by intestinal pathogens. In this way, the gut microflora play a role as symbionts.

INTERACTION BETWEEN MICROBES AND THE RUMEN

Ruminants are animals that ingest and digest cellulose-rich foods and include the herbivores like cow, sheep, giraffe, deer, moose, antelope and goat. These animals do not produce the enzyme cellulase to digest the food, instead they have an organ called the *rumen* where the cellulose is broken down to simpler compounds by the action of rumen microflora and other accessory materials.

The rumen provides a relatively uniform and stable anaerobic environment that has a temperature of 30–40°C, a pH of 5.5–7.0 and a continuous supply of ingested material.

There is a long residence time for the food in the rumen. The rumination process grinds the plant material and provides an increased surface area for microbial attack. The animal's saliva also contributes to rendering the ingested plant material susceptible to microbial attack. The movement of the ruminant stomach supplies sufficient mixing for optimal microbial growth and metabolic activities.

The food in the rumen is mixed by the musculature of the walls of the rumen and is comminuted by chewing when it is returned to the mouth. The liquid part is contributed by the water that is drunk by the animal and the saliva. The saliva contains a buffering system, that makes available all the nutrients in simpler form, i.e. the nutrients that can be digested by the animal is made simpler and is obtained both by the animal and by the microbes in the rumen.

The overall fermentation that occurs within the rumen converts cellulose, starch and other ingested nutrients to CO_2, H_2, CH_4 and low-molecular weight organic acids such as acetic, propionic and butryric acids. The organic acids are absorbed into the bloodstream of the animal, where they are oxidised aerobically to produce energy. The fermentatively produced CO_2 and methane by methanogenic bacteria within the rumen are expelled and do not contribute to the nutrition of the animal.

Microbes of the Rumen

The rumen harbours a great diversity of microorganisms. The bacterial population includes the cellulose digestors, starch digestors, hemicellulose digestors, sugar fermentors, fatty acid utilisers, methanogenic bacteria, proteolytic bacteria and lipolytic bacteria. These populations include

Bacteroides, Ruminococcus, Succinomonas, Methanobacterium, Butyrivibrio, Selenomonas, Succinivibrio, Streptococcus, Eubacterium and *Lactobacillus.*

These bacterial populations produce acetate, the predominant acid within the rumen. The bacteria also produce propionate, the only fermentation acid that can be converted into carbohydrates by the ruminant. Some nitrogen fixation activity also takes place in the rumen.

The next abundant population of microflora is the protozoans. Most are ciliates, but some flagellates such as *Diplodinium, Sarcodina,* etc. are also present. They digest cellulose and starch, some ferment dissolved carbohydrates. Some are predators on bacterial populations. The proteins of the protozoan are in turn digested by the runinant's enzymes. The rumen protozoa store large amounts of carbohydrates which the ruminants digests along with the proteins of the protozoan biomass. The protozoans are digested readily than the bacteria because the latter have resistant cell walls and high nucleic acid contents.

The high diversity of microbial population in the rumen depends on the diet of the ruminant. When there is a sudden change in the diet, there is an upset of the rumen fermentation system resulting in excessive production of methane that can distend the rumen, sometimes to the extent that it compresses the lungs thus suffocating the animal. The fungal population is comparatively low and only some genera of yeasts are present. The food consumed by the ruminant goes to the rumen where there is great microbial diversity and this diversity helps in the digestion of the food consumed. The rumen microbes digest the food for the ruminant and in turn the ruminant provides shelter and food to the microbes.

INSECT–MICROBE SYMBIOSIS

There are other mutualistic relationships in food digestion in the form of intestinal symbionts or as a directed external cultivation of microbial biomass for subsequent consumption. Several insects actually cultivate pure cultures of microbes on plant tissues in a mutualistic relationship.

Various leaf-cutting ant populations maintain mutualistic relationships with fungi. The ants supply leaf tissue for the microbes by inoculating segments of the leaves, and shield the cultivated fungus from competitors maintaining a virtual monoculture that breaks down rapidly if the cultivating ants are removed. For example, Atta species of ants cultivate *Agaricus. Myricocrypta buenzlii* (attine ants) cultivate *Lepiota* (fungal) species.

Attine ants are important because they are responsible for introducing large amounts of organic matter into the soil in rainforests. In the case of the fungus cultivated by attine species of ants the fungus is deficient in proteases and competes poorly with other fungi if not under cultivation by the ants. This particular species of ants chews off sections of green leaves and carries them to a nest where the collected leaves are coated with saliva, cut up into pieces a few millimetres across and wetted with drops of liquid from the anus of the ants. These processes remove or prevent the growth of foreign microorganisms. The prepared leaves are inoculated with pure cultures of fungi (*Lepiota*) which were originally brought by the fertile queen who founded the colony from her home nest. The fungus is grown in pure culture on the leaves even though it is surrounded by soil and the ants bring in many microorganisms from their foraging expeditions. The ants constantly clean their nests, themselves, weed their garden and deposit anal drops onto it. This is done to sterilise their surroundings including themselves and even if any foreign microbe is detected, it is prevented from growing due to the presence of antibiotics in the saliva, and the anal fluid takes care of any that may get through the cleaning.

The fungus is the sole food source of the ants and contains a higher percentage of protein than the original leaves. The fungus supplies cellulase to break down the vegetable matter though the ants add proteases, amylases and chitinases (to lyse any foreign fungi) in the anal fluid. The surprising part of the relationship is that most of the fungi are not known to grow outside ants' nest and the fungi enjoy a competition-free life.

FUNGAL–SCALE INSECT ASSOCIATIONS

This involves the fungal genus *Septobasidium*. Scale insects are plant parasites that become infected with fungi when they emerge from the parent scale. The fungi develop hyphae that surrounds the maturing scale and traps it but does not kill it. The insects live and reproduce within the hyphal mass. The fungal hyphae retain the adult, while the juvenile scale insect feeds on the plant. Here the fungi provide protection to the insect from predators and parasites and in turn, the scale insect provides the fungi with nutrients. The movement of the young scale insect from one plant to another results in the dissemination of the spores and thus the fungus.

The fungus derives its nutrients from the scale insect by hyperparasitism (since the scale insect is itself a parasite on higher plants). The sex of the offspring scale insect is determined by the presence of the fungal spores. If the fungal spores are present the offspring is a female and if not, the offspring becomes a male.

Interaction between Termites and Microbes

Termites are prominent members of arid ecosystems. They are the supreme converters of organic matter in soil from the tropics to the desert. Lignocellulose degradation in termites is mediated by an array of microbes chiefly fungi and bacteria inhabiting the termite intestine. Their associated microbial symbionts dissimilate a significant proportion of the ingested lignocellulose, plant material of which cellulose and hemicellulose are the chief components. The rate of dissimilation could be as high as 85–90%.

Termite Gut Microbes

The evolution of different feeding habits in termites is paralleled by differences in the activity of their gut microbiota. All known termites have a dense and diverse hindgut microbial community which aids in digestion. Termites are divided into two groups — lower and higher termites. The lower termites harbour a dense and diverse population of bacteria and cellulose digesting flagellate protozoa in their alimentary tract. The protozoa belonging to the gut of lower termites represent unique genera and species found nowhere in nature. Higher termites also harbour a dense and diverse array of gut bacteria but they lack protozoa. The bulk of the intestinal microbiota is found in the hindgut (paunch). The termites feed on wood or other vegetation and the bacteria and protozoa degrade this to simple compounds which the termite can use thus enabling it to exploit desert habitats which not many other herbivores can live in.

Studies of gut microflora of *Odontotermus obesus* reveals morphologically diverse bacteria, coccoid and rod-shaped, along with spirochaetes, pseudomonads and actinomycetes. Flagellated protozoans are totally absent. The lower termites (*Pterotermes occidentis*) have a suitable habitat for some microbes in their hindgut or paunch where they set right the problem of digesting cellulose. Some of the main organisms are obligately anaerobic protozoa (flagellates) such as *Trichonympha ampla* in *Pterotermes* and *Mixotricha paradoxa* in some Australian termites.

The flagellated protozoans seem to have cilia but close observations show them to be exosymbiotic spirochaetes (long filamentous bacteria) which are attached to special organelles on the surface of the protozoan and which beat synchronously. There are also other ectosymbiotic bacteria and usually endosymbionts as well. These protozoa use anaerobic bacteria for respiration. Apart from these, there are many other free-living bacteria in the hindgut. There may be up to 10^{10} per ml, either free swimming, attached to the gut wall or to other bacteria or inside other bacteria. Some of these fix atmospheric nitrogen. Hence there are bacteria symbiotic on or in bacteria which are themselves symbiotic on or in the protozoa upon which the termite is dependent.

Synthesis of Metabolites

Besides the hydrolysis of cellulose molecules into glucose monomers, gut microbes play an important role in the synthesis of acetate, CO_2 and H_2 by the fermentation of each glucose monomers. CO_2 reducing acetogenic bacteria convert H_2 and CO_2 to acetic acid which are absorbed from the hindgut and oxidised by respiratory mechanism. This oxidation is aided by methanogenic bacteria.

Hence the guts of both lower and higher termites are a source of microbial diversity. With the isolation of acetogens and methanogens from the guts of termites, the processing of CO_2 and H_2 by the resident microflora is well-documented. It is suggested that termite emissions may be a significant source of atmospheric methane.

Review Questions

1. Explain the various interactions between animals and microbes.
2. Explain the gut and rumen microfloral interactions.
3. Brief out rumen as an ecosystem.
4. Explain the insect–microbe symbiosis.

PART VII

APPLICATIONS OF MICROBIAL INTERACTIONS

Decomposition

Bio-fertilisers

Bio-pesticides

Microbially Induced Corrosion

Bio-mining

Bio-accumulation Xenobiotics

Bio-geochemical Cycles

Nitrogen Fixation

35

Decomposition

Decomposition and photosyntesis are the two important processes of an ecosystem. Litter (the organic remains of biological origin) is an organic chemical carrier of nutrients present in different ecosystems. The structural components of litter govern the rate of decomposition. However, the plant or animal materials present a variety of substances in soil which are heterogeneous both physically and chemically.

Even though agriculture started almost 10,000 years ago, the importance of mineral nutrition for plants came to be recognised only in the nineteenth century.

In 1840, a French scientist Jean-Baptiste Boussingault showed that plants obtain nitrogen from nitrates in the soil. John Bennet Lawes and J. H. Gilbert in England discovered the fertiliser, superphosphate. In 1852, the first inorganic fertiliser, potassium sulphate was produced in Germany. Several biofertilisers have been commercialised for increasing soil fertility and they form interesting examples of microbial interactions which pave the way for increasing the soil fertility.

Significant gaseous components occur in the carbon and nitrogen cycles and to a lesser extent, in the sulphur cycle. Thus a soil, aquatic or marine microorganism often can fix gaseous forms of carbon and nitrogen compounds. In the sedimentary cycles such as that for iron, there is no gaseous component.

Nitrogen, which has stable valence states ranging from –3 as in ammonia to +5 as in nitrate occurs in numerous oxidation states. Nitrogen is a constituent of amino acids, nucleic acids, amino sugars and their polymers. A large, slowly cycled reservoir for nitrogen is nitrogen gas of the atmosphere. Large but essentially unavailable reservoirs of nitrogen are present in igneous and sedimentary rock as bound, non-exchangeable ammonia. Availability of combined nitrogen is an important limiting factor for primary production in many ecosystems.

The inorganic nitrogen ions¾ammonium, nitrite, and nitrate¾occur as salts that are highly water soluble and consequently are distributed in dilute aqueous solution throughout the ecosphere, forming small actively cycled reservoirs.

CHEMICAL NATURE OF ORGANIC MATTER

Soil organic matter comprises the residues of plants and animals at all stages of decomposition mediated by soil microorganisms. Composition of soil organic matter varies with the type of vegetation, nature of soil and its drainage, rainfall, temperature, etc. These components occur in soil in close combination with inorganic substances. They are chiefly made of complex carbohydrates, simple sugars, starch, hemicellulose, pectins, gums, mucilage, protein, fats, oils, waxes, alcohol, aldehydes, organic acids, lignin, hydrocarbons, phenols, etc.

The plant residues may be classified under three major chemical groups—polysaccharides, lignin and proteins. Thus sources of organic matter are forest plants (fallen leaves), dead plants and roots, insects and dead animals.

ORGANIC MATTER DECOMPOSITION

Through a series of complex reactions, microbes mediate the conversion of organic material such as leaves and twigs to the dark humus which colours and glues our soils. Without these processes, the soils would be loose, non-cemented dusts and no life would be able to grow on them, and the world would be a very different place.

Bacteria constitute the most abundant group of microbes, their numbers in soil directly proportional to the amount of organic matter in the soil. Most of the bacteria found in soil are heterotrophs (that which utilise readily available source of organic energy). Some bacteria like *Ferrobacillus* and *Thiobacillus* that make use of inorganic compounds (Fe and S respectively) are not directly involved in organic matter decomposition.

Actinomycetes (mostly thermophiles) too take part in the decomposition process to a great extent by growing on complex substances like chitin and keratin.

Soil fungi are also heterotrophic (mostly) but their numbers vary and depend on whether a species has a dominant vegetative or reproductive phase in the soil environment.

Soil algae vary in numbers and contribute a small amount of organic matter through their biomass but they do not play an active role in decomposition.

The animal groups (earthworms, insects, snails) bring about mechanical reduction by biting and eating the organic matter (pulverising the material). Their excreta enzymes provide good nutrients for bacterial groups to multiply well and decompose the organic matter.

CO_2, H_2O, NO_3, SO_4, CH_4, NH_4 and H_2S are the end products of decomposition. It depends on the availability of air. In more anaerobic environment (stagnant water) one can expect more of H_2S and CH_4.

When we consider the order of decomposition, the first to get attacked are the sugars and starch (water soluble) which undergo rapid attack. Next are the crude proteins followed by hemicellulose, cellulose, oils, fats, wax, resins and lignin.

Organic matter has bound energy which is being taken up by the organisms forming various intermediary substances which are once again attacked by other organisms (which are capable of utilising intermediate products) to liberate CO_2, SO_4, NH_3, etc. Thus the original organic matter is reduced in part to simpler substances which are utilised by various microbes for building up their cell structure. Some of the intermediary breakdown products form solution in soil water and are available for the plants. The undecomposed residues constitute the soil organic complex or humus. In the early stages of decomposition, heterotrophic bacteria are highly active followed by fungi, actinomycetes and other insect groups.

Once organic matter is added to the soil, there is an immediate increase in the heterotrophic microbial population (chiefly bacterial). Their life span is very short. After initial decomposition of the organic matter, they reach a stage of optimum growth and finally lyse. The lysed cells release simpler substances (partly taken up by the plants). The insoluble constituents form part of the humus. Thus organic matter added to the soil is converted by oxidative decomposition to simpler substances which are made available in stages for plant growth and the residue is transformed into humus. This process is called *humification*.

During organic matter decomposition, sugars, starch, hemicellulose and cellulose are rapidly broken down into CO_2, H_2O and some energy. Oils, fats, waxes and resins are slowly decomposed to give CO_2 and H_2O. Lignin is resistant to decomposition in that some portions undergo slow decomposition and the end products are usually aromatic compounds (compounds containing benzene ring). The other portion chemically unite to become a part of soil humus.

Proteins are broken down slowly (since the clay minerals trap the protein molecules making it unavailable to the microbes). This problem is evaded by some microbes that excrete extracellular proteases, but since the enzymes themselves are proteinaceous, they lose their activity substantially. Apart from this difficulty, proteins also form an organic complex in soil with the lignin and become resistant to microbial attack, which converts it into amino acids, amides and ammonium which is oxidised to nitrites and nitrates which is again taken up by the plants.

Phosphorus in phytin, nucleic acids and cell membrane are broken down to orthophosphates. Similarly, sulphur containing amino acids are decomposed to liberate sulphates.

Three processes go on during decomposition:

1. Degradation of plant and animal remains (mineralisation) by cellulases and other enzymes.

2. Increase in biomass of microbes (polysaccharides and proteins) (immobilisation)

3. Accumulation or liberation of end products (index of microbial activity in soil).

C : N Ratio

The amount of carbon and nitrogen present in the soil depicts the C : N ratio (C : N ratio of humus is 10 : 1) and this is very critical for the decomposition process. Optimum levels of C : N ratio (80 : 1–12 : 1) is ideal for maximum decomposition since a favourable soil environment is created to bring about an equilibrium between mineralisation and immobilisation process which will be upset if C : N ratio is less than the optimum. For example, if the carbon content in the fresh organic matter is higher when compared to the nitrogen content (i.e. if the C : N ratio is high), then there will be more multiplication of microbes and immobilisation of the available nitrogen thus making it unavailable to the plants. On the other hand, if the fresh organic matter contains more of nitrogen than the optimum, then it will result in accumulation of the mineralised nitrogen in the form of ammonium and nitrates thus increasing toxicity of the environment (since carbon is relatively less, the multiplication of the microbial cells is hindered due to lack of energy).

Towards the end of the decomposition process, C : N ratio becomes small (due to the cessation of microbial multiplication leading to no intake of carbon for energy and lysis of the microbial cells thus resulting in release of bound nitrogen). Due to accumulation of nitrogen, the ratio may reach 10 : 1. Hence whatever be the initial C : N ratio during the early stages of decomposition, it always comes to lie at 10 : 1 towards the end thus stabilising the decomposition process.

Benefits of Organic Matter Decomposition

- There is improved seed germination (humic substances) and root growth, and uptake of minerals by plants (mobilisation of NPK from the soil into the root) is increased in the presence of humic substances.

- Uptake of trace elements by plants is increased since humus is known to chelate with trace elements like iron.

- Humus enhances the enzyme activity involved in plant metabolism.

- Humic acids are known to influence the growth and proliferation of microbes. For example, growth of *Aspergillus*, *Penicillium*, *Bacillus* is enhanced by addition of humus. The number of *Azotobacter* cells and amount of nitrogen fixed by them are enhanced by application of humic acids.

- Organic matter contributes to the exchange capacity of cations and anions in the soil, and affects the retention, release and availability of plant nutrients.

- It binds organic chemicals and pesticides which affects their biological activity and toxicity. It lessens some of the harmful effects of chemicals dumped into the environment and improves water percolation and retention into the soil.

- It is involved in the formation and maintenance of desirable soil structure.

- It absorbs solar radiation which influences soil temperature. It provides colour to soil and decreases its reflectance.

- The more organic matter a soil has, the darker the soil will be and the more energy it will absorb.

Anaerobic Decomposition of Organic Matter

Decomposition of organic matter takes place by activity of mesophilic and thermophilic microbes thus producing CO_2, H_2, C_2H_5OH and organic acids like acetic acid, formic acid, lactic acid, succinic acid and butyric acid. Among the mesophilic flora, bacteria (*Clostridium*) are more active than fungi or actinomycetes in cellulolytic activity. They are found in large numbers in peaty and manured soils but rarely found in cultivated arable soils.

In compost heaps, both mesophilic and thermophilic microbes (bacteria and actinomycetes) are important in the breakdown of cellulose substrates. The initial stages of organic matter decomposition is by the breakdown of complex carbohydrates and proteins into organic acids and alcohols. The later stage envisages the methane bacteria which are strict anaerobes and which act on secondary substrates chiefly lactic acid, acetic acid and butyric acid and ferment them into ethyl alcohol and CO_2. Examples of methane bacteria are *Methanobacterium, Methanobacillus, Methanosarcina* and *Methanococcus*.

SOIL HUMUS

Soil humus is a mixture of dark, colloidal organic compounds relatively resistant to decomposition. These compounds result from the decay of organic litter, and accumulate in the O and A horizons of soils. Soil humus helps to glue mineral particles into aggregates, giving structure to the soil and affecting soil stability.

Humus is formed during the decomposition of organic 'litter' (including pine needles, leaves, and animal droppings) in soils. This decay is mediated by microbes and the enzymes they excrete (which break certain specific bonds in organic matter). The main reactions of this decomposition are:

aerobic conditions: Carbohydrate $+ O_2 \rightarrow CO_2 + H_2O +$ Energy

anaerobic conditions: Carbohydrate $\rightarrow CO_2 +$ Acid or alcohol + Energy

Humus includes sugar amines, nucleic acids, phospholipids, vitamins, sulpholipids, polysaccharides and many other unclassified compounds. Fulvic acid and humin have similar structures. The COOH and phenolic OH groups are weakly acidic, which give humus its pH buffering ability, pH dependent charge and cation chelating ability. In addition, the toxicity of the phenolic subgroups which make up humus contributes to its resistance to microbial decomposition. The branched structure of humic molecules may also make them more resistant.

There are three main classifications of humus: fulvic acid, humic acid and humin. Humin is insoluble but fulvic and humic acids are soluble in dilute NaOH solution. Humic acids precipitate in acidic solution, but fulvic acids remain soluble. Humic molecules are incredibly varied in composition, but generally are characterised by:

- Many active chemical functional groups exposed to the surrounding solution for reaction with other substances in the solution.

- A very large cross-linked and folded molecule with molecular weights in hundreds of thousands of grams per molecule.

Humus has a large surface area per unit of mass and is highly charged (similar to clay), and individual humus molecules are dynamic and constantly changing form (but may remain as humus for several thousand years).

These reactions are much complicated by the complex structure of the litter, and the products of decomposition include various nutrients, organic acids and amines (depending on the conditions and starting materials) in addition to the resistant residues or humus. Between 60 and 80 percent of the carbon from most plant residues is evolved as CO_2 within a year of deposition; 5 to 15 percent is incorporated into the microbial biomass and the rest remains in soil humus.

There are different pathways of decomposition of organic matter to form humus. In one of the pathways, polyphenols are produced (either from the decomposition of lignin—a rigid polymer that, with cellulose, makes up woody material—or through microbial synthesis from nonlignin sources), then enzymatically oxidised to form quinones which polymerise creating humic macromolecules. Early in decomposition, simple, phenolic compounds, bits of lignin and melanin (lignin-like molecules) and other phenolic polymers are transformed by beta-oxidation of side chains, addition of hydroxyl groups, oxidation of methyl groups and decarboxylation. The more reactive compounds are oxidised and then converted to radicals, which stabilise by linking into dimers or forming quinones (oxidised polyphenols). These linkages repeat to create humic macromolecules.

During the decomposition or humification of organic litter, the carbon to nitrogen (C/N) and carbon to sulphur (C/S) ratios decrease, indicating that relatively more carbon than nitrogen or sulphur is lost in the process.

Numerous factors control or influence the decomposition of organic matter. Among these are the properties, amount and stage of decay of the organic matter, and the availability of oxygen, temperature, soil moisture, nutrients and soil texture. Temperature and water availability affect the decomposition rate. As acidity increases, soil respiration decreases, leading to an accumulation of organic matter and a drop in the rate of litter decomposition. When soil moisture is considered, the presence of live roots stimulate the decomposition of organic material because the roots increase microbial activity.

Review Questions

1. Define litter.
2. Describe in detail the decomposition of organic matter.
3. What are the end products of decomposition?
4. Define humification.
5. What is C : N ratio? Explain.
6. What are the benefits of organic matter decomposition?
7. Discuss the anaerobic decomposition of organic matter.
8. What is soil humus?
9. What is humin?

36

Bio-fertilisers

Soil possesses the inherent capacity to support growth of plants by supplying proper amount of different nutrients in the proper proportion. This capacity depends upon the amount of plant nutrients present in the soil.

Fertilisers are essentially inorganic chemical substances (either used singly as in the case of urea or in combination as in NPK) that supply nutrients to the growing plants in the right proportion so as to enable proper growth of plants. The advent of fertilisers was a boon to the farmers who had solely depended on farm yard manure and rotation of crops to restore soil fertility.

TYPES OF FERTILISERS

There are principally two types of fertilisers namely chemical and biological. In India, the Green Revolution brought in its wake, massive use of chemical fertilisers leading to contamination of water by nitrates and increased salinity of the soil. Chemical fertilisers are very expensive from the energy point of view leading to high cost and dependence on nonrenewable energy sources. For example, Factamphos and NPK are chemical fertilisers.

Chemical fertilisers have the following effects on soil.

- short-term increase of soil fertility
- immediate increase in soil fertility
- non-biodegradable residues are left in the soil leading to biomagnification
- chemicals are leached into water

To counteract the above disadvantages of chemical fertilisers, the focus has therefore shifted to bio-fertilisers which are advantageous in the following ways.

- they improve soil structure
- they are long-term fertilisers
- no residues are left in soil
- gradual distribution of the required nutrient is seen
- act as biocontrol agents

BIO-FERTILISERS

Bio-fertilisers are preparations containing active or latent cells of efficient strains of certain microbes that can utilise the atmospheric nitrogen to increase the nitrogen content of soil, and can dissolve the insoluble phosphate of the soil to release the phosphorus it contains in the soluble form for increasing crop yield. Bio-fertilizers are of the following types:

- bacterial
- fungal
- algal
- aquatic fern
- earthworms

BACTERIAL BIO-FERTILISERS

Here, live cells of bacteria are used as fertilisers. They are the principal nitrogen fixers in the soil and may be either free-living or symbiotic (within the nodules of legume roots, leaf nodules of certain plants and stem nodules of *Sesbania grandiflora*). Some bacterial genera can solubilise phosphorus from the bound form (*Bacillus, Pseudomonas* sp.). There are certain other bacterial members which during their metabolic activity, make available certain essential trace elements like manganese, calcium and zinc in the soil for plant absorption. Some bacteria like *Agrobacterium* sp. and *Pseudomonas fluorescence* supply growth hormones like indole acetic acid to the plants. Certain groups of bacteria like the *Pseudomonas fluorescence* living in association with the rhizosphere of most of crop plants (rhizobacteria promoting plant growth) supply all the essential nutrients required for the growth of the crop and in addition, protects the plant roots from the attack by soil-borne pathogens (saprophytic suppression). The nitrogen fixing ability of the bacteria can be studied under the following heads.

Free-living nitrogen fixers   Among the large number of free-living nitrogen fixers, only two have attracted the scientists, they are *Azotobacter* and *Klebsiella*. *Azotobacter* requires oxygen to flourish and *Klebsiella* and *Rhodospirillum* can survive both in the presence and absence of oxygen. Apart from fixing nitrogen, *Azotobacter chroococcum* has the ability to synthesise and secrete B-vitamins, growth hormones and antifungal antibiotics into its environment. *Azotobacter* has one drawback, its nitrogen fixing ability is regulated by the presence of nitrogenous compounds in its environment. Thus it cannot be used along with a nitrogenous fertiliser. *Klebsiella* is another promising free-living nitrogen fixer, and it has been possible to isolate its 'nif' genes (17 of them) and to introduce them into *E.coli*. It is hoped that one day it will be possible to introduce these genes into non-nitrogen-fixing plants to make them self-sufficient in nitrogen.

Symbiotic nitrogen fixers   *Rhizobium* is a soil inhabitant, and grows along the leguminous roots, invades it and forms nodules within which it transforms itself to bacteroids when it starts fixing atmospheric nitrogen directly into the roots of the plant. As early as 1895, scientists introduced a laboratory grown culture of *Rhizobium* in the trade name of *Nitragin*. The culture contained nodule extracts, gelatin, sugar and asparagine in a solid medium. Now, pure cultures have replaced the nodule extract (7 different species) and *Rhizobium* is sold in two different forms: one as an inoculant to be used during seed sowing and second as pelleted seed. The first attempt at genetic engineering of *Rhizobium* was to improve its performance. Scientists in USA have engineered *Rhizobium meliloti* to produce more nitrogen. They have cloned the 'nif' gene from the *Rhizobium* and re-inserted a number of copies of this gene into the bacterium. The engineered strain increased the crop productivity by about 17%.

Isolation of Rhizobium   The leguminous plants are uprooted. Their root nodules are carefully detached from the roots and washed well with sterile water. The nodules are sterilised with 0.1% $HgCl_2$ for 5 minutes and then washed with alcohol and sterile water (6 times). The prepared nodules are taken in a test tube and crushed, adding a little sterile water. This inoculum is transferred to a nitrogen free growth medium containing mannitol. After 48 hours of incubation, a loopful is transferred to solid media (yeast extract mannitol agar [YEMA]). Colonies of *Rhizobium* are evident as shining water drops after a further incubation of 24–48 hrs. Thus the bacterium is isolated.

Preparation of seed inoculant Legume seeds are taken (*Trifolium* sp.). The seeds are surface sterilised with 0.1% $HgCl_2$ followed by a wash with alcohol and several sterile water washes. A thick slurry is made out of jaggery with water. The seeds as well as the culture scraped from the solid medium are added to the cooled sticky slurry and stirred well. Jaggery acts as a binder of cells to the seeds so that better adherence is achieved. Inoculated seeds are shade dried and used for sowing. Thus a seed inoculant with Rhizobium is prepared.

There have been many failed attempts to introduce 'nif' genes into monocots. One of the reasons for this failure is that cereals do not have root hairs, thus they offer a barrier to the entry of bacteria. Scientists succeeded in initiating nodule formation in cereals (by treating the root tip with cellulase to dissolve the cell wall and exposing the same to culture of *Rhizobium* in the presence of polyethylene glycol which dissolves the cell membrane of the cell and allows the bacteria to enter). But the nodules did not fix nitrogen.

There is another bacterium that helps cereal plants to fix nitrogen to some extent. It is *Azospirillum* which inhabits both root cells as well as the surrounding of the roots (rhizosphere) forming an associative symbiotic relation and increasing the nitrogen fixing potential of the cereal plants. Methods have been developed to use it as a bio-fertiliser since when pelleted, the bacterium remains viable up to 31 weeks. In transplanted rice, seedlings dipped in a slurry of *Azospirillum* inoculum showed substantial fixation of nitrogen.

Drawbacks

- Regular use of *Azospirillum* for nitrogen fixation results in breaking down of nitrates to nitrites by the bacterium and then further action on nitrites to form atmospheric nitrogen, i.e. they aid in denitrification process.

- Since plants absorb nitrogen mainly as nitrates, to make *Azospirillum* a viable fertiliser, mutants have to be obtained which do not have the two deleterious effects.

Phosphatic Bio-fertilisers

The soil is inhabited by a group of bacteria termed as phosphobacteria which have the capacity of releasing bound phosphates in the soil and thus making it available for the plants. They act in different ways. Some bacteria secrete organic acids such as lactic acid, acetic acid and citric acid which solubilise the bound phosphate to forms which are available to plants. Others produce sulphuric acid by oxidation of sulphur which acts like the organic acids in making phosphate available. Carbonic acid formed by the action of carbon dioxide released by bacteria during respiration and water acts in a similar way.

Some organisms give off hydrogen sulphide which reacts with the iron salt, ferrous phosphate, to form ferrous sulphide and thus releases the phosphate.

Microbes are also responsible for decay of dead animal and plant litter in the soil during which humic and fulvic acids are formed which bind the metal ions such as Fe, Al, Mn and Ca and release the phosphate ions for the plants.

The most common varieties of phosphobacteria are *Pseudomonas* species and *Bacillus megaterium*. Like Rhizobium, they are used as seed inoculants.

Algae as Bio-fertilisers

Another group of free-living nitrogen fixers are the cyanobacteria commonly called the blue-green algae (BGA). More than a hundred species of BGA can fix nitrogen.

Nitrogen fixation takes place in specialised cells called the heterocysts (large, thick walled and metabolically inactive cells) which depend on vegetative cells for energy to fix nitrogen while the fixed nitrogen is utilised by the vegetative cells for growth and development. BGA are very common in the rice fields (the micro-aerophilic condition and alkalinity are conducive to the algal population). If no chemical fertilisers are added, inoculation of the algae can result in 10–14% increase in crop yields. Unlike *Azotobacter*, the BGA are not inhibited by the presence of chemical fertilisers.

On the other hand, presence of inorganic minerals (superphosphate) accelerates their growth. They are easy to produce and usually they are mass produced in cement tanks filled with fresh-water. Since they do not require any processing, they are quite cheap. Cost of 10 kg may be only Rs.30–40. Algal effect can be seen in subsequent years without the use of fresh inoculants. Examples of some algal bio-fertilisers are *Anabaena*, *Nostoc* and *Oscillatoria*. Algae have proved beneficial in the case of certain crops like vegetables, cotton and sugarcane when added along with chemical fertilisers.

Aquatic Fern (Azolla) as a Bio-fertiliser

Azolla is a tiny water fern common in ponds, ditches and rice fields. It has been used as a bio-fertiliser for rice in all major rice growing countries including India, Thailand, Korea, Philippines, Brazil and West Africa. The nitrogen accumulated in the *Azolla* is made available to the rice crop when the fern decomposes. The nitrogen fixing work is accomplished by the symbiotic relationship between the fern and a BGA, *Anabaena azollae*. The alga inhabits some of the cells on the underside of the *Azolla* frond and fixes atmospheric nitrogen. It is dependent on the fern for photosynthates which supply the energy for nitrogen fixation. In addition to nitrogen, the decomposed Azolla also provides K, P, Zn and Fe to the crop. It also controls aquatic weeds which would otherwise compete with the crop for nutrients.

Azolla and rice productivity *Azolla* provides at least 100,000 tonnes of nitrogen fertiliser/year worth more than $50 million annually. *Azolla* can increase rice yields as much as 158% per year. Rice can be grown year after year, with no decline in productivity, hence no rotation of crops is necessary.

Cultivation Cultivation of *Azolla* is easy, and it can be grown on any open tank though it may be affected by seasonal fluctuations. Studies are on in China to produce new varieties of *Azolla* that will withstand the changing climatic conditions. People have maintained buckets of *Azolla* in ordinary tap water. Use of Azolla may be an important factor in the world's future food needs and may play an important role in reducing the world's reliance on fossil fuel-based fertilisers. The significance of its symbiotic relationship with *Anabaena* is astounding when one considers that millions of lives depend on these two organisms.

Actinomycetes as Bio-fertiliser

Frankia is an actinomycete and forms nitrogen fixing nodules in trees and shrubs. The organism invades the cells of a developed lateral root and causes it to fuse into a nodule. Entry into the host changes the structure of the microbe. Scientists are hopeful that some day they may be able to make fruit trees like apple, pear, plum, raspberry, etc. by fixing nitrogen through the involvement of *Frankia*.

Fungi as Bio-fertiliser

Some nonpathogenic fungi help in plant growth by forming associations with the host plant roots called mycorrhizae (*myco–* fungi, *rhiza* –root). Some examples of such fungi are *Trichoderma, Gigaspora, Glomus,* etc. One group of mycorrhizae forms a sheath around the fine lateral roots and replaces the root hairs by dichotomous branching of the fungal hyphae. They are called *ectomycorrhizae* because they do not traverse intracellularly. The ectomycorrhizae help the plant by solubilising nutrients near the plant roots and making it easy for the plants to feed. They also prevent the roots from being attacked by nematodes (by entangling them). Another group called the *endomycorrhizae* penetrate the roots and establish symbiotic relation with the plants. The fungi helps the roots in obtaining inorganic nutrients while obtaining essential organic nutrients from the host. There is yet another group called ect-endomycorrhiza or vesicular-arbuscular mycorrhiza (VAM fungi) wherein they are partly outside the host roots and partly intracellular.

The mycorrhizae are of particular significance in agriculture as they assist plants to absorb phosphates from the soil. Phosphate ions are not very mobile in the soil, thus when a plant uses up all the available phosphate near its roots, it starves for want of phosphates. The mycorrhizal mycelia spread far and wide, away from the depleted zone and translocates the phosphates directly to the host. Plants with restricted root system, having short stubby roots and few root hairs are particularly benefited by association with mycorrhiza. Absorption of elements which are required in minute quantities by plants are also facilitated by mycorrhizae. Mycorrhizae are valuable particularly in times of drought. They enable plants to survive and grow in the driest and poorest of soils, even in leached soils that have been mined by deforestation and overcropping.

Cultivation of mycorrhizal fungi It is easy to mass cultivate ectomycorrhizal fungi since they can grow in an artificial fungal medium. They are mass produced in liquid media, dried and pressed into cakes or cultivated in the form of spores and packaged. The seed inoculation is performed just as that of *Rhizobium*. Cultivation of endomycorrhizal fungi is quite difficult since they need a live host for survival. Hence the host plants harbouring such mycorrhizal fungi are specially grown in laboratory conditions, their roots excised, powdered and the mixture used as a fertiliser.

Earthworms as Bio-fertiliser

Earthworms are farmers' friends. They dig up and mix the soil, eat up decayed plants and convert them to fertiliser thus enriching the soil. In recent years a low-tech biotechnology has emerged to restore earthworms to their natural place in the environment through vermiculture (since the continuous use of chemicals have caused their depletion from soil). Vermiculture requires mineral inputs in terms of ingredients (leaf litter, household and agricultural wastes along with a starting population of earthworms). Prof.Ismail of the Department of Zoology, New College, Chennai is an authority in vermiculture and is producing vermicompost by allowing earthworms instead of microorganisms in the soil to convert the waste plant matter into compost.

Review Questions

1. Define bio-fertilisers.
2. List the benefits of bio-fertilisers.
3. Explain the various types of bio-fertilisers.
4. How do you isolate rhizobium and use it as a seed inoculant?

37

Bio-pesticides

A fascinating and economically important area of applied microbial ecology is the biological control of pests and disease-causing agents using microorganisms. Biological control using microorganisms to initiate disease in pest populations has the potential of reducing agricultural reliance on chemical pesticides. The negative interactions (amensalism, predation, and parasitism) among microbial populations and between microbes and higher organisms have always formed a natural basis for the biological control of pests and pathogens, and biological control methods take advantage of those relationships. Biological methods to control populations of disease-causing organisms and pests are based on modification of host and/or vector population, modification of reservoirs of pathogens, and the direct use of microbial pathogens and predators. Biotechnology may improve approaches for controlling pest and disease causing populations.

BIOLOGICAL CONTROL

Biological control is the reduction of inoculum density or disease producing activities of a pathogen or parasite in its active or dormant state by one or more organisms. This is accomplished naturally or through manipulation of the environment/host/antagonist or by mass introduction of one or more antagonists.

Principles of Biocontrol

Biocontrol is based on antagonism. Which is further based on competition (for nutrients and space), parasitism (by production of volatile/non-volatile antibiotics) and hyperparasitism.

- *Competition*   In the soil, the dormant propagules o7f fungi are subjected to intense microbial competition when nutrients become available, e.g. Ammonium sulphate solution improves the biocontrol of *Heterobasidium annosum* by *Trichoderma viridae.*

- *Hyperparasitism*   It is the attack of a secondary parasite on primary parasite (parasitism of one microbe by another) observed among airborne pathogens. Rusts are often parasitised by *Enderluca carices* and powdery mildews by *Ampelomyces quisqualis. Cladosporium* species parasitise both rust and powdery mildews.

- *Antibiosis*   It depends on competitive saprophytic ability of a fungus. Sterile soil does not produce antibiotic substances. Several phylloplane fungi such as *Aureobasidium, Alternaria, Botrytis* and *Helminthosporium* have been shown to produce antibiotics.

MICROBIAL INSECTICIDES

Pests and diseases can be effectively controlled without chemical pesticides by good crop management practices.

Integrated Pest Management

It is a law that emphasises the need to restrict insect population rather than to eradicate them so that the pest population is kept at a non-injurious level. IPM is a pest management system that in the context of the associated environment and the population dynamics of the pest species utilises all suitable techniques and methods in as compatible a manner as possible to maintain the pest population at levels below those causing economic injury.

Population of pathogenic or predatory microorganisms that are antagonistic towards a particular pest population provide a natural means of controlling pest population and preparations of such antagonistic microbial populations are called *microbial pesticides* or *microbial insecticides.*

An effective microbial pesticide possesses the following characteristics:

- The microbial pathogen must be virulent and cause disease in the pest population when properly applied at the recommended concentration.

- Pathogen should not be susceptible to environmental conditions.

- After application, it should survive long until the infection is established within the pest population. Resistant stages of the pathogen like spores or cysts are best for use.

- The pathogen should be rather specific for the pest population and must not cause disease in the non-pest population.

- It should rapidly establish disease in the pest population so as to minimise destruction caused by the pest.

- The microbial pesticide should be harmless to humans and other valued plant and animal population.

The various types of microbial insecticides include (a) bacterial pesticides (b) viral pesticides and (c) fungal pesticides.

BACTERIAL PESTICIDES

Several bacterial pathogens that have been used as insecticides include (a) endospore forming *Bacillus* and *Clostridium* species (b) non-endospore forming species of *Pseudomonas, Enterobacter, Proteus, Serratia, Xenorhabdus*. Of the potential bacterial pesticides, *Bacillus thuringiensis* has been most extensively studied.

Bacillus thuringiensis has been successfully tested against more than 140 insect species (*Lepidoptera, Hymenoptera, Diptera* and *Coleoptera*). At present, there are 12 groups of *B. thuringiensis*. All strains produce protein crystal inclusion bodies which act as endotoxin. They are toxic factors and are called *parasporal bodies*.

These crystals dissolve under alkaline condition. They are not soluble in water under neutral or acidic condition. The midgut contents of the caterpillar larvae (pests) are alkaline. On ingestion, the crystal dissolves in the midgut fluid and gets digested particularly by the proteolytic enzyme present in the midgut fluid. This digested protein crystal attacks the cementing substances which are present in the gut wall, thus loosening the epithelial gut wall which helps continuous diffusion of liquid from the gut into the blood making the blood of the insect highly alkaline and leading to gut paralysis. The parasporal bodies are highly toxic for caterpillars with an LD_{50} value of < 0.9 μg/g of larvae.

The process of crystal synthesis and spore formation proceed simultaneously. The toxin production by a culture can be enhanced by controlling many factors.

Bacillus thuringiensis toxin genes have been introduced via recombinant DNA technology into the genome of plants or plant associated microorganisms.

Method of Production

Submerged fermentation

Preparation of inoculum The inoculum is prepared using media containing beet molasses, corn steep solids and calcium carbonate.

Production medium Production medium enhances the production of toxins within the cell which happens during sporulation. The fermentable carbohydrate and nitrogen available for growth are exhausted at the same time after the commencement of sporulation.

Submerged fermentations have been mainly used to produce flowable formulations of spore crystal complex to be used as *sprays.*

Semisolid fermentation Medium consists of dextrose, yeast autolysate and a potassium source. This medium is used for propagating the culture in flasks. The production medium to produce spores and toxins contain wheat bran and solid substrate along with other nutrients. This is an aerobic fermentation with an incubation period of 36–48 hours. This fermentation gives a wettable powder or dust form which is got by just drying and grinding the bran cake. This product is more stable than flowable one but it cannot be used in the wet form since bran begins to swell when wetted.

Bacillus thuringiensis in the Field

Insecticides based on *B. thuringiensis* are used for short-term control. They do not persist from year to year. During crop spraying, it is essential to make sufficient coverage on a leaf surface. But the *B. thuringiensis* spray must be protected from inactivation by sunlight since it is sensitive to ultra-violet light. In order to avoid this, emulsifying agents, stickers or binders and UV protectants can be used in the formulation. Some commercial names of *B.thuringiensis* are DIPEL, DOOM.

Bacillus popilliae and *B. lentimorbus* are potent insecticides causing *Milky disease* in Japanese beetle. Following the ingestion of viable spores by larvae, germination takes place in the gut and the vegetative cells penetrate the gut wall and enter the body cavity where they multiply. During this time, they remove all the nutrients from the blood and lead to the death of the larva.

VIRUSES AS INSECT PESTS

Insect viruses are encased in protein coats which are insoluble in water. These protein crystals serve to protect the virus particles and due to the presence of these protein crystals, the infective viruses remain active even under normal storage condition for many years.

Specificity of the virus–host relationship makes viruses an ideal candidate for use in the control of specific pest population. There are two major groups of insect viruses:

(i) Polyhedrosis virus where many virus particles are embedded in each protein crystal. When they occur in the host cell nucleus, they are called nuclear polyhedrosis virus (NPV), and when they occur in the host cytoplasm they are called cytoplasmic polyhedrosis virus (CPV).

(ii) Granulosis virus where only one virus is contained in each protein crystal. They either develop in the nucleus or cytoplasm of the host cells (NGV or CGV). NPV and CGV come under the group of Baculoviruses and they are the most extensively studied insect viruses.

Infection is caused by ingestion of contaminated food containing NPV/ GV followed by cell invasion beginning in the midgut. The GV are sprayed over the leaves and they enter the larvae feeding on the foliage. The viruses multiply fast in the larvae and make them lethargic. Affected larvae become sluggish and stop feeding and die. For example, NPV controls the pest *Helicoverpa armigera*.

The affected larvae hang upside down from the leaves and twigs in a characteristic way and a brownish fluid oozes from them. This is a highly infective fluid and is readily disseminated amongst the healthy insect population. The spread takes place by wind, rain, etc.

TAIVIRIDAE is a commercial NPV pesticide used to control cotton pest bollworm and budworm.

Production

The specificity and obligate parasitism of insect viruses makes their mass production possible only on live insects, e.g. production of bollworm-budworm NPV (*Baculovirus helothis*).

Bollworm larvae are raised on a semi-synthetic diet containing a water-based mixture of casein, sucrose, wheat germ, growth factors, etc. Agar is used to solidify the diet. Chemicals like formalin, sorbic acid, aureomycin are used to inhibit bacteria, yeasts and fungi. The diet is dispersed hot from a large tank to plastic trays and sealed. These are inoculated by caterpillars from an insectary and incubated under controlled conditions to produce a mass of larvae. A known volume of virus is sprayed on the diet which replicates on the caterpillars to the extent of 5000–10000 times in 5–7 days. The infected caterpillars are suctioned and then treated with water, filtered, centrifuged, precipitated and spray dried. After quality control testing, the preparation is packaged and sold.

FUNGAL PESTICIDES

Insect mycoses are caused by Phycomycetes, Ascomycetes, Basidiomycetes and Deuteromycetes, e.g. *Beauvaria, Metarrhizium, Entomophthora, Hirsutella.*

Biological Control Agents

Trichoderma is a cosmopolitan biological control agent which is a saprophyte.

- It has rapid growth rate.
- It sporulates abundantly.
- It produces antibiotics (gliotoxin and viridian).
- It is a well known antagonist.

Commercial product BINABTSEEPIC can be easily applied in soil in the form of granules for control of soil-borne diseases.

Some examples of fungal agents   *Trichoderma harzianum* is successfully used as a bio-control agent for diseases caused by *Rhizoctonia solanii* on beans and tomato. *T. harzianum* is the most active antagonistic agent in the case of *Fusarium* species.

Control of Airborne Diseases

Botrytis cinerea was controlled when treated with *Trichoderma* species on the senescent (aged) floral parts. Their colonisation prevented the saprophytic establishment of *B. cinerea* on the vine (hyperparasitism). Their establishment on the floral parts limit the development of disease on the branches. Spore suspension of *Trichoderma* sp decrease the number of rust pustules caused by *Puccinia.*

Seed-borne diseases   *T. harzianum* and *T. koningii* when applied to pea seeds resulted in reduction in the incidence of pre-emergence of damping off of the pea.

Fruit diseases   *Botrytis rot* of strawberry was controlled with *Trichoderma*. *Anthracnose* by *Colletotrichum* was also effectively controlled by *Trichoderma* sp.

Mechanism of Control

Antibiotics may play a crucial role in microbial antagonism. Lytic enzymes like chitinase and B -1-3, glucanase also help in controlling the diseases.

Mode of Application

Mycelial preparation of *T. harzianum* is more effective in comparison to spore preparation of biological control caused by *Sclerotinia* spp. Wheat bran/peat is colonised by *Trichoderma* sp.

A liquid medium is used to inoculate the bran/corn meal along with other nutrient ingredients like glucose, sodium chloride, calcium carbonate and soybean meal.

Semi-solid fermentation is preferable for most of the fungi. Since the dry conidia bran mixture is relatively stable. Embedding propagules of the biocontrol agents in a matrix formed by sodium alginate (binder) and clay appear to be an effective method of delivery of *T. viridae*. It can also be applied as a seed coating with various adhesives/gels. Some commercial products available are ANTAGON, BIOCURE-F, BIODERMA, TRICHOCAN, etc. Those commercial formulations can be introduced into organic manures, composts, dry cow dung, farmyard manure, neem cake etc.

ENTOMOPATHOGENIC NEMATODES

Entomopathogenic nematodes belonging to the family Steinernematidae and Heterorhabditidae have emerged as biocontrol agents. Both have a third stage infective juvenile (IJ) which is a non-feeding larval stage and adapted for long-term survival in the soil. They are symbiotically associated with their respective bacterium, *Steinernema* with *Xenorhabdus* and *Heterorhabda* with *Photorhabdus* which causes rapid death of the insect host.

Each nematode species has a specific association with one bacterial species. IJ of certain species actively seek out for their host and exhibit a strategy like 'hunter'. Others wait for their prey and are designated as 'ambushers'. Nematodes of both the genera are easy to produce on large scale in liquid medium and are safe to humans.

In general, the IJ that are present in the soil are always in resting stage. When an insect host is present in the soil, the IJ moves towards the host.

Firstly, they find the host. Secondly, they penetrate into the suitable hosts through natural openings; by punching the host cuticle and finally, the IJs have to overcome the insect defence system and start multiplying.

The IJs release the symbiotic bacteria and the insects die 2–4 days after infection. Nematodes are able to persist for a long time in the environment. Persistence is the measure of the number of live nematodes present in the soil, e.g. *Steinernema scapterisei* are infective to mole crickets.

GENETIC ENGINEERING IN BIOLOGICAL CONTROL

Genetics has always played an important role in biological control, but the deliberate release of genetically engineered microorganisms is scientifically controversial because of possible undesirable ecological effects.

Transgenic plants have been engineered to increase their defences against microbial pests and pathogens. For example, the genes for phytoalexins, which are produced by varieties of soybean, pea and tomato in response to microbial infections, have been introduced into transgenic plants to protect them against various fungal infections.

Review Questions

1. What is biological control?
2. Write a note on principles of biological control.
3. What are microbial insecticides? Write some of their characteristics.
4. Write in detail about various microbes used as pesticides.

38

Microbially Induced Corrosion

Bio-corrosion is one of the direct consequences of microbial film formation on the surface of water distribution pipes. It is one of the major contributor to water quality and environmental contamination. Bio-corrosion causes severe economic losses in water distribution systems.

Corrosion of iron and steel pipes can occur as a result of variety of chemical reactions that establish an electrochemical gradient, leading to loss of metal from the pipe due to electrolysis. The physical presence of microbial cells on a metal surface, as well as their metabolic activities, can cause Microbiologically Influenced Corrosion (MIC) or bio-corrosion. The forms of corrosion caused by bacteria are not unique. Bio-corrosion results in pitting, crevice corrosion, selective de-alloying, stress corrosion cracking, and under-deposit corrosion. Biofilms provide the localised environmental conditions (e.g. decreased pH; differential oxygen cells) for initiating or propagating corrosion activities.

The metabolic capabilities of microorganisms are being harnessed to improve the recovery of metals and petroleum from the environment. Sulphur-oxidising thiobacilli are commercially employed in bioleaching operations for the recovery of copper and uranium. Microorganisms play both beneficial and detrimental roles in the mining and mineral processing of metals.

CAUSES OF CORROSION

Corrosion is caused by any one or more of the following mechanisms.

Oxygen Influencing Corrosion

Non-uniform (patchy) colonies of biofilm result in the formation of differential aeration cells where areas under respiring colonies are depleted of oxygen relative to surrounding non-colonised areas. Having different oxygen concentrations at two locations on a metal causes a difference in electrical potential and consequently corrosion currents. Under aerobic conditions, the areas under the respiring colonies become anodic and the surrounding areas become cathodic. Oxygen depletion at the surface of stainless steel can destroy the protective passive film. Since stainless steels rely on a stable oxide film to provide corrosion resistance, corrosion occurs when the oxide film is damaged or oxygen is kept from the metal surface by microorganisms in a biofilm. MIC-associated bacteria are grouped on the basis of their mode of attack on ferrous and non-ferrous metals. The most common MIC groups include sulphate-reducing, iron-oxidising, acid-producing, sulphur-oxidising and nitrate-reducing bacteria. Acid production, hydrogen sulphide generation, tubercle formation and the subsequent development of differential aeration cells can lead to deterioration and failure of mild steel, copper, stainless steel, and other ferrous and non-ferrous metals used as construction materials.

Oxygen depletion at the surface also provides a condition for anaerobic organisms like sulphate-reducing bacteria (SRB) (Fig. 38.1) to grow. This group of bacteria are one of the most frequent causes for bio-corrosion. The metabolic activities of anaerobic sulphate-reducing bacteria result in the formation of iron hydroxides which are corrosion products.

Sulphur bacteria obtain energy by reducing or oxidising inorganic sulphur compounds that are present in feed waters. The bacteria most often associated with MIC in water systems belong to the anaerobic sulphate-reducing (SRB) group, which includes *Desulfovibrio desulphuricans*.

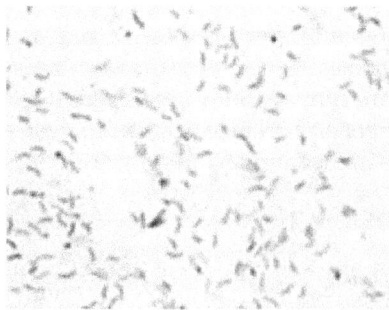

Figure 38.1 *Sulphate reducing bacteria*

Direct attack of ferrous and non-ferrous metals by their hydrogen sulphide metabolic by-product is a significant problem in many industries. Reduction of sulphate to H_2S (addition of electrons) results in cathodic depolarisation. Sulphate reducing bacteria accelerate the electrolytic corrosion process by promoting depolarisation of the anodic (+) and cathodic (-) surface during the anaerobic corrosive reaction. H_2S reacts with ferrous ion to convert it to ferrous sulphide—effect of this reaction is anodic depolarisation. Additionally, a very active hydrogenase associated with *Desulfovibrio* species removes the protective layer of hydrogen that surrounds submerged iron pipes, exposing the underlying iron to corrosive attack.

Aerobic bacteria near the outer surface of the biofilm consume oxygen and create a suitable habitat for the sulphate-reducing bacteria at the metal surface. Sulphur oxidising bacteria, such as *Thiobacillus* species, are aerobic microorganisms that can produce sulphuric acid. This group of organisms often lives in close association with SRB. SRBs can grow in water trapped in stagnant areas, such as dead legs of piping. Symptoms of SRB-influenced corrosion are hydrogen sulphide (rotten egg) odour, blackening of waters, and black deposits. The black deposit is primarily iron sulphide.

Nitrate Reducing Bacteria

Nitrate reducing bacteria (NRB) can utilise nitrogen containing organic compounds in feed waters, producing significant quantities of ammonia. In addition to odour problems, ammonia production is associated with stress corrosion cracking of copper alloys. Nitrite-based corrosion inhibitors may be a source of nitrogen for this group of MIC bacteria.

Acid-producing Bacteria

Bacteria can produce aggressive metabolites such as organic or inorganic acids. For example, *Thiobacillus thiooxidans* produces sulphuric acid and *Clostridium aceticum* produces acetic acid. Acids produced by bacteria accelerate corrosion by dissolving oxides (the passive film) from the metal surface and accelerating the cathodic reaction rate.

Hydrogen-producing Bacteria

Many microorganisms produce hydrogen gas as a product of carbohydrate fermentation. Hydrogen gas can diffuse into metals and cause hydrogen embrittlement.

Iron Bacteria

Iron-oxidising bacteria obtain energy through oxidation of reduced ferrous species to the ferric state. Iron oxidation by bacterial species in this group usually results in the formation of ferric hydroxide, $Fe(OH)_3$, which is precipitated in their slime.

Iron-oxidising bacteria, such as *Gallionella*, *Sphaerotilus*, *Leptothrix*, and *Crenothrix* are aerobic and filamentous bacteria which oxidise iron from a soluble ferrous (Fe^{2+}) form to an insoluble ferric (Fe^{3+}) form. The dissolved ferrous iron could be from either the incoming water supply or the metal surface. The ferric iron these bacteria produce can attract chloride ions and produce ferric chloride deposits which can attack austenitic stainless steel. For iron bacteria on austenitic stainless steel, the deposits are typically brown or red-brown mounds.

Anaerobic Microbial Corrosion

This type of corrosion of cast iron causes graphitisation, a process in which a pipe loses much of its iron thereby becoming soft and brittle. Steel and aluminium pipes are also subjected to anaerobic corrosion. Anaerobic microbial corrosion of steel results in localised pitting which sometimes causes perforation of the pipe.

Pitting Corrosion

Pitting corrosion (Fig. 38.2) is a localised form of corrosion; the bulk of the surface remains unattacked. Pitting is often found in situations where resistance against general corrosion is conferred by passive surface films. Localised pitting attack is found where these passive films have broken down. Pitting attack induced by microbial activity, such as sulphate reducing bacteria (SRB) also deserves special mention. Within the pits, an extremely corrosive micro-environment tends to be established, which may bear little resemblance to the bulk corrosive environment. For example, in the pitting of stainless steels in chloride-containing water, a micro-environment essentially representing hydrochloric acid may be established within the pits. The pH within the pits tends to be lowered significantly, together with an increase in chloride ion concentration, as a result of the electrochemical pitting mechanism reactions in such systems.

The detection and meaningful monitoring of pitting corrosion usually represents a major challenge. Pitting failures can occur unexpectedly, and with minimal overall metal loss. Furthermore, the pits may be hidden under surface deposits, and/or corrosion products.

Monitoring pitting corrosion can be further complicated by a distinction between the initiation and propagation phases of pitting processes. The highly sensitive electrochemical noise technique may provide early warning of imminent damage by characteristic signals in the pit initiation phase. Figs. 38.2 (a, b) show the extent of pitting corrosion.

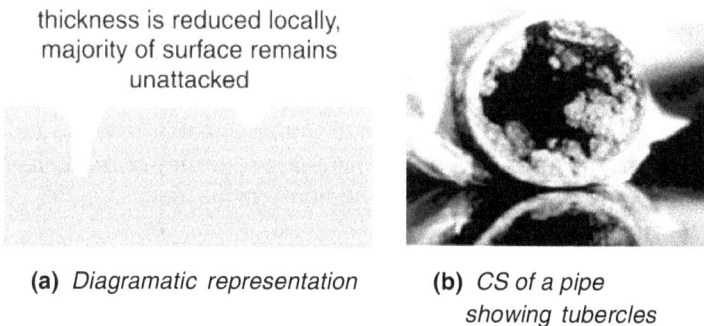

thickness is reduced locally, majority of surface remains unattacked

(a) *Diagramatic representation*

(b) *CS of a pipe showing tubercles*

Figure 38.2 *Pitting Corrosion*

Pipe failures resulting from microbiologically influenced corrosion (MIC) have been widely recognised in petrochemical, gas and nuclear power industries, but only recently has this phenomenon been associated with failures in fire protection systems (FPS). MIC results in mechanical blockages of piping and sprinkler heads, as well as through-wall penetration of ferrous and non-ferrous metals. FPS are designed for the life of the structures in which they reside; however, reports of new systems developing MIC-associated through-wall leaks within months of installation are becoming more prevalent.

Pitting corrosion occurring under deposits in FPS can be initiated or propagated by these microbial activities. Through-wall penetration of carbon steel and copper has been reported within months after a new pipeline has been brought into service. This extensive tuberculation can cause occlusion of pipelines, sometimes completely blocking flow in six-inch diameter pipelines. These problems become more critical as pipe diameter decreases, posing a potential threat to proper sprinkler head mechanical functioning. In addition, FPS make-up waters are typically stagnant, soft (relatively low in hardness), acidic and devoid of antimicrobial agents such as the sodium hypochlorite that is used for microbial control in potable waters. These characteristics predispose FPS to biological fouling and MIC. Regulatory requirements that dictate periodic testing can also contribute to development of MIC in FPS when make-up waters are replaced with oxygenated and nutrient-rich waters. MIC-associated microorganisms can use these nutrients as growth sources, leading to fouling of affected systems.

The most serious consequence of MIC in FPS is mechanical blockage of piping and sprinkler heads. MIC-associated organisms can attach to the metallic surfaces of FPS, forming corrosion deposits that are termed tubercles (as shown in Fig 1.20). Tubercles can completely occlude pipes, and more significantly, these deposits can break off and block sprinkler head flow channels. Localised pitting-type attack can also occur underneath tubercles, resulting in through-wall penetration. The resulting acid production, hydrogen sulphide generation and development of differential aeration cells can lead to the loss of essential metallic properties of mild steel, copper, stainless steel and other ferrous and non-ferrous metals.

Protection from Corrosion

Pipes can be protected from corrosion by following the procedures given

- By increasing the pH to 9.5 pipelines can be protected against the action of sulphate reducing microbes.

- Buried pipes can be coated to prevent contact between metal surface, water and soil microbes.

- Electric currents can be applied to the pipe to preclude corrosion processes.

- Various bacterial inhibitors can be employed to control microbial corrosion. For example, alkyl substituted amine and quaternary ammonium compounds are toxic to microbes. Various bacterial inhibitors can be employed to control microbial corrosion. For instance, alkyl substituted amine and quaternary ammonium compounds are toxic to many bacteria like *Desulfovibrio* sp., a bacterium of major importance in the corrosion process.

MICROBIALLY INDUCED CONCRETE CORROSION

Microbially Induced Concrete Corrosion is an important biological or chemical phenomenon that is having extreme effects on the infrastructure of our cities. We are conducting research that is designed to provide more insight into the biochemical and chemical reactions occuring, the microbial ecology of concrete corrosion as well as to allow us to develop process based models of concrete corrosion and develop control mechanisms to prevent or control concrete corrosion. It is found that aerobic heterotrophs, and neutrophilic and acidophilic sulphur oxidisers are the dominant microbes. There are also SRB, anaerobic heterotrophs, nitrate reducing bacteria, and ammonia oxidising bacteria present in some of the samples.

The corrosion of concrete pipes is a consequence of a cyclic process caused by microbial sulphur metabolism. Two types of sulphur metabolism are involved in the cycle of sulphur in the environment. One is an anaerobic process in which H_2S is produced by anaerobic bacteria, the other is an aerobic process in which the H_2S is oxidised to elemental sulphur (S) or sulphuric acid (H_2SO_4). This cyclic process exists as a natural method for the cycling of sulphur compounds in the environment and may also exist in sewage collection systems.

During the transport of raw sewage from the top of the sewage collection system to the treatment plants, the organisms in the sewage start to degrade the abundant organic compounds present in the raw sewage. This often results in a depletion of O_2 from the sewage. This results in the creation of anaerobic or anoxic conditions which allow the growth of sulphate reducing bacteria (SRB) which grow only in the absence of O_2 and obtain energy by utilising small organic compounds or H_2 as energy sources and transferring the electrons produced to sulphate, thus reducing it to sulphide.

The sulphide produced eventually partitions into HS⁻ and H_2S. The H_2S is a gas and evolves into the headspace of the sewer pipes, reaching the crown of the pipe. The crown of the pipe is exposed to an aerobic environment which supports the growth of sulphur oxidising bacteria. The sulphur oxidising bacteria grow on and within the concrete of the pipe, oxidising the H_2S present and producing H_2SO_4. The sulphuric acid dissolves the CaOH and $CaCO_3$ in the cement binder, thus causing corrosion of the concrete pipes.

There have been only a few species of thiobacilli (the largest genera of organisms that oxidise H_2S to H_2SO_4) described by researchers. These are *T. novellus*, *T. thioparus*, *T. neopolitanus*, *T. intermedius* and *T. thiooxidans*. The first four organisms are important for establishing the acidic conditions necessary for corrosion to occur, while the acid loving *T. thiooxidans* grows in conditions of very low pH and produces H_2SO_4 in copious amounts, thus lowering the pH even more.

Review Questions

1. What is bio-corrosion?
2. What are the causes of corrosion?
3. Write a note on anaerobic microbial corrosion.
4. Write a short note on pitting corrosion.
5. How can pipes be protected from corrosion.
6. Write a note on microbially induced concrete corrosion.

39

Bio-mining

Microorganisms degrade certain toxic constituents used in mineral processing and concentrate and immobilise soluble heavy metals released as a result of mining and mineral processing activities.

Certain bacteria are responsible for one of the most persistent and destructive environmental problems—acid rock drainage (ARD).

The microbial ecology of mining environments is far from being fully elucidated. Most microbial research and development efforts related to metal mining environments have been directed towards the bacteria involved in acid rock formation and the commercial applications of bioleaching and mineral bio-oxidation. The reason for this is two fold: ARD is the most widespread, persistent, destructive and least controllable environmental problem associated with base and precious metal mining. But the same bacteria that cause ARD can be effectively applied in controlled situations to process base and precious metal deposits that are not technically and economically amenable to treatment by conventional mineral processing methods.

MICROBIOLOGY AND CHEMISTRY OF ARD

Acid Rock Drainage is the leachate resulting from the oxidation of sulphide minerals exposed to water, air, and bacteria and the resultant products from the interaction of acid, metal-bearing solutions reacting with alkaline rocks and water. ARD is the result of both bacterial and chemical activity.

When sulphide minerals (pyrites) are exposed to air and water in rocks during mining operations, the pyrite chemically oxidises, creating a slightly acidic environment conducive for the development of *Thiobacillus ferrooxidans* which colonises the exposed mineral surfaces. It derives energy from the oxidation of inorganic sulphur and iron-containing compounds.

Bacterial oxidation of pyrite produces ferric iron, a strong oxidant that chemically oxidises mineral sulphides including pyrite. The ferrous iron resulting from this reaction is regenerated to ferric iron by *T. ferrooxidans*. The ferric iron is then available to oxidise more pyrite, and the cycle continues. Apart from *T. ferroxidans* are the chemoautotrophic acidophilic *Thiobacillus thiooxidans*, which oxidises reduced sulphur and *Leptospirillum ferroxidans*, which oxidises reduced iron compounds are also present in the same environment.

The oxidation of pyrite is an exothermic reaction. In waste rock piles with active pyrite oxidation, sufficient heat is produced and retained within the pile for heat to build up. It is not unusual for temperatures to exceed 60°C. Because *Thiobacillus* and *Leptospirillum* species are mesophilic, oxidising inorganic substrates in the temperature range of 10–40°C, the high temperatures that occur in some sulphidic waste piles eventually limit these organisms. At lower temperatures (10–40°C) *Thiobacillus* and *Leptospirillum* species will predominate. At about 40°C, mesophilic chemoautotrophs will begin to die and moderately thermophilic, acidophilic chemoautotrophic bacteria will appear. At 30–40°C, mesophilic and moderately thermophilic bacteria will coexist. Oxidising iron and sulphur compounds at a temperature range of about 40–60°C is a diverse group of organisms that are not yet well characterised. At a temperature of approximately 55°C, the moderate thermophiles are succeeded by the extremely thermophilic, acidophilic *Sulfolobus, Metallosphaera, Sulfobacillus* and *Sulphurococcus archaea*. These organisms oxidise iron and sulphur compounds under acid conditions at temperatures ranging from about 55–85°C. When the temperatures exceeds the upper limits of the bacteria colonising the moist and acidic areas of the sulphidic rock pile, all bacterial activity ceases and the temperature of the pile will decrease.

Because of the bacterial catalysis in ARD formation, efforts have been made to inhibit the growth of *Thiobacillus* and *Leptospirillum* species and the thermophilic bacteria by adding surfactants and slow release biocides. But frequent applications are required. Once ARD is initiated, it is virtually unstoppable. To prevent the initiation of ARD, sulphide minerals must be isolated from air and water. This stops chemical oxidation and also inhibits the growth and activity of the bacteria that catalyse the reactions accelerating formation of ARD.

BIOLEACHING AND BIO-OXIDATION

As destructive and as unstoppable as ARD is the catalytic activity of the mesophilic and thermophilic chemoautotrophs that has been harnessed for cost-effective, efficient and environmentally acceptable commercial processing technologies called bioleaching and mineral bio-oxidation. Bioleaching is the bacterial oxidation of sulphide minerals, whereby metals of value (copper, uranium and zinc) are released into solution. Mineral bio-oxidation is a biological process in which iron sulphide minerals such as pyrite and arsenopyrite are degraded by bacteria and precious metals (gold and silver) are liberated for recovery by conventional metallurgical techniques.

Since 1950, bacteria have been used to bioleach sulphidic, mineral-bearing waste rock from surface copper mines. The waste rock is piled near mining operation, a dilute sulphuric acid solution is applied to the top surfaces of the waste piles by using drip irrigation, and the solution is percolated through the pile. The moist, acid environment along with air entering from the tops and sides of the waste rock pile provides a conducive environment for naturally occurring *Thiobacillus* and *Leptospirillum* species to develop. Bacterial numbers reach 10^6 to 10^7 per ml of each solution. The bacteria oxidise copper sulphide minerals releasing soluble copper and ferric iron which is carried from the waste pile by the percolating acid solution. The copper is recovered by solvent extraction, and high grade cathode copper is produced by electrowinning. Some 20% of the world's copper is estimated to be produced by bioleaching. Even higher grade copper sulphide ores are bioleached in a process called bacterial thin layer leaching. The copper sulphide ore is crushed to less than 6.3 mm and placed on an impermeable pad to a height of 3–6 m. A dilute sulphuric acid solution is applied along with a mixed culture of mesophilic iron and sulphide oxidising bacteria, and bacterial catalysis commences. After 7–9 months, bioleaching is complete, with about 80% of the copper extracted from the ore.

A heap leaching method has been developed to process precious metal ores in which elemental gold and silver are encased in sulphide minerals. Precious metal ores, amenable to mineral bio-oxidation, are called refractory sulphidic precious metal ores. As with bacterial leaching of copper sulphide ores, the refractory sulphidic precious metal ores are crushed but usually to a larger size and stacked on impermeable pads to heights of approximately 6–12 m. Bacteria can be added to the crushed ore as it is stacked onto lined pads. With drip irrigation of the heap using dilute sulphuric acid containing low concentrations of ammonium and phosphate ions, the bacteria in close proximity to the sulphide minerals rapidly oxidise the pyrite and arsenopyrite in which the gold and silver are embedded. Within several months, depending on the ore characteristics, the bacteria and ferric iron have oxidised the sulphide minerals, exposing the elemental gold and silver. The ore is then washed with water to remove acid, soluble heavy metals and iron. The ore is neutralised and treated with a dilute sodium cyanide solution or other reagents (thiourea or thiosulphate) that solubilise the precious metals. Bio-oxidation of refractory sulphidic precious metal ores is called pretreatment because the ore is subjected to an additional treatment process before undergoing conventional metallurgical extraction with gold solubilising reagents.

Review Questions

1. What is rock drainage?
2. Describe the process of bio-mining.
3. What is bio-leaching?
4. Describe the heap leaching method of bio-leaching.

40

Bio-accumulation

Various heavy metals like mercury, cobalt, tin, nickel, cadmium and thallium are used in metal alloys or as catalysts, and during their mining and ultimate disposal, cause heavy metal pollution problems. All these metals are substantially toxic to plants, animals and many microorganisms.

Microorganisms, owing to their large surface to volume ratio and high metabolic activity, are important vectors in introducing heavy metal and radionuclide pollutants into food webs. From neutral to alkaline pH, heavy metals in soils and sediments tend to be immobilised by precipitation and/or adsorption to cation exchange sites of clay minerals. Microbial production of acid and chelating agents can reverse this adsorption and mobilise the toxic metals. Microbial metabolic products that can chelate metals include dicarboxylic and tricarboxylic acids, pyrocatechol, aromatic hydroxyl acids, polyols and some specific chelators such as enterochelins and ferrioxamines.

Mobilisation is often followed by uptake and intracellular accumulation of the heavy metals, both by microorganisms and by plant roots. It is not entirely clear why some of these toxic metals are taken up and stored by microorganisms, but intracellular sequestering seems to confer heavy metal resistance on at least some bacteria. Filamentous fungi were shown to transport heavy metals and radionuclides along their hyphae. This has some implication for the potential role of mycorrhizal fungi in transmitting such pollutants into higher plants. Direct root uptake of heavy metals mobilised by microbial acid production or chelation is an alternative possibility.

The heavy metal cadmium is of special concern in this respect. Cadmium is highly toxic and tends to accumulate with even very low exposures because it is excreted extremely slowly. Its approximate half-life in humans is ten years. Cadmium in humans causes, at low chronic exposure, hypertension and kidney damage. Higher exposures through rice grown on industrially contaminated fields have caused the painful and crippling bone and joint disease (itai-itai disease). Cadmium occurs in low concentrations in phosphate fertiliser spread on agricultural fields. Another potential source of cadmium and other heavy metals is sewage sludge. Such sludge is being considered for use as soil conditioner. As a substantial portion of sewage sludge is derived from microbial biomass, the relatively high concentration of heavy metals in this material also reflects the ability of microorganisms to concentrate these pollutants.

Microbial accumulation of radionuclides, have clear human health implications. Lichens are extremely effective in concentrating the radionuclides such as ^{90}Sr and ^{137}Cs from atmospheric fallout. During periods of snow cover, when the lichens are shielded from direct fallout, concentrations of radionuclides in the lichens decrease. During periods of rain, concentrations of radionuclides increase. Because lichens serve as the primary producers in a food chain of lichen-caribou-humans, there can be an efficient transfer of such concentrated elements to the highest member of the food chain. These radionulcides, being deposited in bone tissue, may affect blood cell synthesis in the bone marrow and cause leukemia.

Review Questions

1. Define bio-accumulation and describe its characteristics.

41

Xenobiotics

Xenobiotics are chemical compounds synthesised by humans which are not naturally found in living organisms and cannot normally be metabolised by them.

The xenobiotic compounds have molecular structures and chemical bond sequences not recognised by existing degradative enzymes. These resist biodegradation or are metabolised incompletely, with the result that some xenobiotic compounds accumulate in the environment. There are many reasons for a xenobiotic organic compound proving recalcitrant (totally resistant to biodegradation):

1. Unusual bonds or bond sequences (as in tertiary and quaternary compounds).

2. Unusual substitutions (as with chlorine and other halogens).

3. Highly condensed aromatic rings.

4. Excessive molecular size (as in polyethylene and other plastics).

5. Failure of a compound to induce synthesis of degrading enzymes.

6. Failure of the compound to enter the microbial cell for lack of suitable permeases.

7. Unavailability of the compound due to insolubility or adsorption phenomena.

8. Excessive toxicity of the parent compound or its metabolic products.

The term *biodegradation* has been used to describe transformations of every type, including those that yield products more complex than the starting material as well as those responsible for the complete oxidation of organic compounds to CO_2, H_2O, NO_3 and other inorganic components. Sometimes, the microbial transformations result in residues that are more stable than the parent compound, yet this phenomenon is called degradation because the parent compound disappears.

The distribution of recalcitrant compounds tends to dilute in the environment but still it is a cause of concern because of the phenomenon called biomagnification. To be subject to this phenomenon, the pollutant must be both persistent and lipophilic, as a result of which minute dissolved amounts of these substances are partitioned from the surrounding water into the lipids of both prokaryotic and eukaryotic microorganisms. Concentrations in their cells, compared to the surrounding medium, may increase by one to three orders of magnitude. Members of the next higher trophic level then ingest the microorganisms. Only 10–15% of the biomass is transferred to the higher trophic level, the rest dissipated in respiration but the persistent lipophilic pollutant is neither degraded nor excreted to a significant extent, and so is preserved practically without loss in the smaller biomass of the second trophic level. As a result, its concentration increases by almost an order of magnitude. The same thing occurs at successively higher trophic levels. The top trophic level, composed of birds of prey, mammalian carnivores and large predatory fish, may carry a body burden of the environmental pollutant that exceeds the environmental concentration by a factor of 10^4–10^6.

If the pollutant is a biologically active substance, such as a pesticide, at such levels it may cause death or serious debilitation of the affected organism. Chlorinated hydrocarbons, including DDT are implicated in the death or reproductive failure of various birds.

Many xenobiotic pollutants that have proven recalcitrant to microbial attack are halocarbons. The carbon-halogen bond is highly stable. Cleavage of this bond is not exothermic but rather requires a substantial energy input; it is an endothermic reaction. As a result, halocarbons are chemically and biologically very stable. Important groups of halocarbons include the solvents and refrigerants, haloaromatics such as chlorobenzenes, chlorophenols and chlorobenzoates, polychlorinated or polybrominated biphenyls and triphenyls, chlorodibenzodioxins and chlorodibenzofurans. Some organochlorine insecticides are also highly recalcitrant.

Review Questions

1. What are xenobiotics?
2. What are the reasons for any compound to be recalcitrant?
3. What is biodegradation?
4. Define bio-magnification?

42

Bio-geochemical Cycles

Microorganisms, in the course of their growth and metabolism, interact with each other in the cycling of nutrients, including carbon, sulphur, nitrogen, phosphorus, iron and other trace elements. This nutrient cycling called bio-geochemical cycling, when applied to the environment, involves both biological and chemical processes. Nutrients are transformed and cycled, often by oxidation–reduction reactions that can change the chemical and physical characteristics of the nutrients. All of the bio-geochemical cycles are linked and the metabolism related transformations of these nutrients have global-level impacts.

CARBON CYCLE

Carbon cycle is the driving force behind nearly all nutrient cycling reactions involving organic sulphur, nitrogen or phosphorus. Any biologically synthesised organic compound can be decomposed by soil microbes. This theory was first proposed by a famous soil microbiologist Martin Alexander.

If this were not true, then there would be vast accumulation of undecomposed carbon compounds. Any organic compound that contains energy in reduced bonds is ultimately used as an energy source. The carbon cycle is the greatest natural recycler of carbon atoms.

Sources of Carbon

Carbon dioxide which is predominantly present in air is also found in the soil and oceans (dissolved CO_2).

Most of the earth's carbon is locked up in fossil fuels as carbonates. 10% of CO_2 is released through industrial and human activities like burning of fossil fuels, agricultural activities and forest clearing and burning. CO_2 is also released by soil microbial respiration. Table 42.1 gives the sources of carbon.

Table 42.1 *Various sources of carbon*

Sink	Amount in billions of metric tons
Atmosphere	578 (as of 1700) to 766 (as of 1999)
Soil organic matter	1500 to 1600
Ocean	38,000 to 40,000
Marine sediments and sedimentary rocks	66,000,000 to 100,000,000
Terrestrial plants	540 to 610
Fossil fuel deposits	4000

The carbon cycle has two parts:

Slow cycle Here the carbon turnover is measured in 100s and 1000s of years involving weathering of rocks and dissolution of carbonates on land and oceans.

Fast cycle Carbon turnover measured in years or in decades. This is biological in nature. This cycle directly affects and is affected by soil microbes. A schematic representation of the carbon cycle is shown in Fig. 42.1.

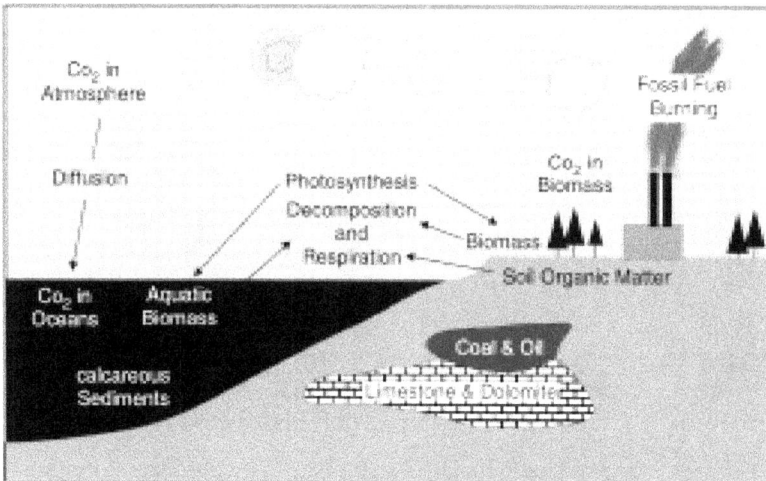

Figure 42.1 *Schematic representation of the carbon cycle*

Sources of Organic Carbon

Carbohydrates   These are the most common types of organic carbon entering the soil. The most abundant organic compound on earth is cellulose, which provides the primary structural component for plants. (Chitin, present in insects, crustaceans, and bones, is the second most abundant organic compound.) Like starch, cellulose is a polymer of glucose monomer units, linked together at the beta-1,4 locations as opposed to the alpha-1,4 locations for amylose (insoluble starch). Enzymes are generally extremely specific in their catalytic actions. They can recognise even the subtlest difference in the substrate structure and often exhibit no measurable catalytic behaviour toward other similarly structured substrates. The difference in the glucose linkage between starch and cellulose makes it impossible for the starch digesting enzymes, e.g. alpha-amylase, to break down cellulose. The direct consequence of this specificity is that various organisms, including humans, cannot use cellulose to satisfy their nutritional requirement for carbohydrates.

However, some animals and insects, such as cattle, sheep, horses, termites, and caterpillars, can subsist on wood and grass, although they themselves do not produce cellulolytic enzymes. This is due to the synergistic effect of the bacteria present in their digestive tracts. These gut bacterial flora secrete the necessary cellulolytic enzymes to digest cellulose, and the hosts, in turn, provide them with a shelter as well as nutrients.

Sugars   Monosaccharides are simple sugars which are building blocks and universal energy substrates for soil microbes. Disachharides are present mostly in fungal cell wall materials.

Lignin   It is second to cellulose in terms of biomass. It is the structural component of plant which imparts rigidity and resistance to compression, bending and pathogens. It protects cellulose and hemicellulose from enzymatic attack. It is a cementing agent, amorphous and highly branched with no defined structure. It is a complex polymer of coumaryl, coniferyl subunits. It is an aromatic ring (phenyl) with a three-carbon side chain (phenyl propanoid). Lignin is the richest source of aromatic (carbon) compounds in nature.

Fats, waxes and hydrocarbons   They are carbon compounds soluble in ether. Fatty acids are components of cell membranes. They are either saturated or unsaturated (contain one or more double bonds between carbon atoms).

Some of the problems associated with microbial utilisation of plant carbon include physical barriers by cell wall compounds, cutins, suberins, chitins etc. lignin which slows the rates of decomposition by microbes, and the microbicidal properties of some plants.

Water is required for hydrolysis and to increase surface area for enzymatic degradations.

Plants contain a mixture of chemical components like cellulose, hemicellulose, pectin, lignin, waxes and their decomposition requires combined action of many microbes, none of which have all the enzymes required to completely decompose the whole plant material. Any environmental factor affecting the soil biota affects the decomposition also. For example, decomposition of organic matter is rapid in tropical regions when compared to temperate since the temperature there is warmer and microbes have a high metabolic rate. Substrate quality has an effect on decomposition which depends on C : N ratio, if C : N ratio is greater than 20-30 : 1, then immobilisation of nitrogen results and consequently decomposition of carbon slows down. For example, lignin content slows down the rate of decomposition because it has no nitrogen.

Mineralisation of Carbon

Carbon metabolism depends on whether microbes need carbon containing compounds or other nutrients like nitrogen, phosphorus or sulphur. Metabolism results in mineralisation. Organic carbon is converted into carbon dioxide and release of inorganic minerals like ammonia, phosphates, sulphates takes place. The obvious sign of mineralisation of carbon is respiration in soil.

Decomposition and Carbon Dioxide Evolution

By this phenomenon, the limited supply of CO_2 available for photosynthesis is replenished. Three separate simultaneous processes can be distinguished during organic carbon transformation:

- Plant and animal tissue constituents disappear under the influence of microbial enzymes.

- New microbial cells are synthesised so that the proteins, polysaccharides and nucleic acids typical of bacteria and fungi appear.

- Certain end products of the breakdown are excreted into the surroundings to accumulate or to be further metabolised.

Carbon Assimilation

Organic carbon decomposition serves two functions for the microflora – providing energy for the growth and supplying carbon for the formation of new cell materials.

Cells of microbes contain 50% carbon. The source of the element is the substrate being utilised. The process of converting substrate to protoplasmic carbon (organic carbon) is called assimilation.

During aerobic metabolism, 20–40% of substrate carbon is assimilated and the remainder is released as carbon dioxide. Fungal flora releases less carbon dioxide than other microbial groups because fungi are more efficient in their metabolism (converting substrate carbon to cell carbon since fungi have a complex cell structure). Hence in fungi 30–40% carbon metabolised is used to form new mycelium. Aerobic bacteria show only 5–10% of carbon assimilation whereas among anaerobes, utilisation of carbohydrates is very inefficient, about 2–5%. At the same time, as carbon is assimilated for the generation of new cells, there is a simultaneous uptake of N, P, K and S thus immobilising these nutrients.

NITROGEN CYCLE

Plants, animals and most microbes require combined forms of nitrogen for incorporation into cellular biomass but the ability to fix atmospheric nitrogen is restricted to a limited number of bacteria, archaea, and symbiotic associations.

1. Uptake of NH_4 or NO_3 by organisms

2. Release of NH_4 by decomposition

3.4. Microbial oxidation of NH_4 (yields energy in aerobic conditions)

5. Denitrification (NO_3 respiration) by microbes in anaerobic conditions (NO_3 is used instead O_2 as the terminal electron acceptor during decomposition of organic matter)

6. Nitrogen fixation

7. Nitrate leaching from soil

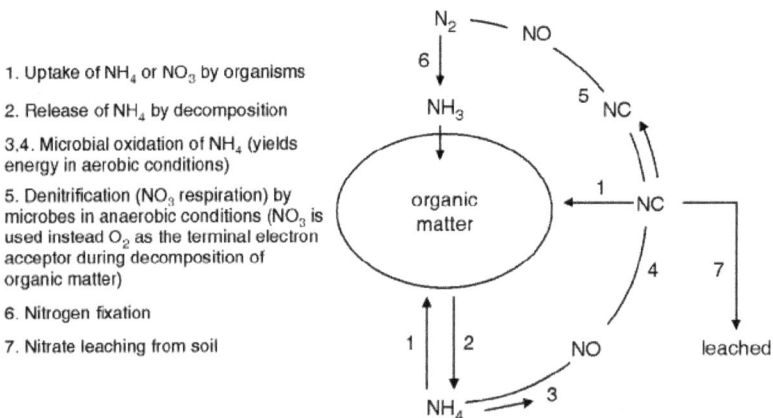

Figure 42.2 *Schematic representation of nitrogen cycle*

The bio-geochemical cycling of the element nitrogen is highly dependent on the activities of microorganisms.

The positive associations of di-nitrogen fixing bacteria with certain plants provide essential combined nitrogen for crops and ecosystems. One of the most important mutualistic relationships between microorganisms and plants involved the invasion of the roots of suitable host plants by nitrogen fixing bacteria, resulting in the formation of a tumour-like growth called a nodule. Within the nodule the nitrogen fixing bacteria are able to convert atmospheric nitrogen to ammonia, which supplies the nitrogen required for bacterial and plant growth.

Sources of Nitrogen

- Atmosphere contains about 79%.
- Soil and rocks contain bound non-exchangeable ammonia which are released slowly through biological weathering.
- Inorganic nitrogen ions such as ammonia, nitrites and nitrates occur as salts that are water-soluble and distribute themselves in the soil solution.
- Living and dead organic matter also form small, actively cycled reservoirs of nitrogen, e.g. humus.
- Fertilisers.
- Volcanic activity, ionising radiation and electrical discharges supply additional combined form of nitrogen to the atmosphere to be added to the soil.

Steps in Nitrogen Cycle

During the cycling of nitrogen, two major processes namely mineralisation and immobilisation of nitrogen occur.

Mineralisation   Here the bound organic unavailable form of nitrogen (proteins, nucleic acids, amides, amines, urea, etc.) get converted to easily available and assimilable form of nitrogen (ammonia, nitrite and nitrate) by the action of microbes present in the environment (terrestrial and aquatic). This process occurs as ammonification and nitrification processes.

Immobilisation   The available form of nitrogen (ammonia or nitrate) are being utilised by plants (at large) and by the microbes themselves (for their cellular needs) and thus the inorganic form of nitrogen gets converted into bound organic form. In another case, the atmospheric nitrogen (gaseous form) both during nitrogen fixation and denitrification gets fixed inside the microbial cell as utilisable form (proteins) wherein it is called nitrogen fixation. The three primary processes involved in the N cycle are nitrogen fixation, nitrification, and denitrification. In nitrogen fixation, N_2 serves as a nitrogen source for bacteria. In nitrification, ammonia or nitrite serves as the source of reductant and energy for chemolithotrophic growth. In denitrification, N oxides serve as terminal electron acceptors to support microbial respiration, usually in the absence of O_2.

AMMONIFICATION

It is an important step in the nitrogen cycle that helps in the mineralisation of bound nitrogen (organic). This is performed by a large number of organisms present in the soil that play a role in decomposition.

Ammonification is thus the mineralisation of bound organic nitrogen into inorganic ammonia.

Sources of Nitrogen for Ammonification

- protein
- amines
- amides
- urea
- nucleic acid (bases)

A number of bacteria, fungi and actinomycetes attack the organic nitrogen at varying rates of decomposition.

As a first step, the fungi attack the proteins breaking them down to polypeptides. Then the bacteria act on the polypeptides and convert them to aminoacids which are then cleaved to release ammonia.

Proteins $\rightarrow \rightarrow \rightarrow$ Polypeptides $\rightarrow \rightarrow$ Aminoacids

The bacteria that help in ammonification include *Pseudomonas, Bacillus, Clostridium* and *Serratia*, while the fungi genera include *Alternaria, Mucor, Aspergillus* and *Penicillium*.

- Microbes synthesise extracellular proteolytic enzymes for protein decomposition.

- Major end products of aerobic proteolysis are carbon dioxide, ammonia, sulphates and water.

- Anaerobic decomposition of proteins (putrefaction) releases foul- smelling compounds like H_2S, mercaptans, etc.

- Fungi release less ammonia than bacteria since they assimilate more of the nitrogen for cellular synthesis (since fungi have a greater cellular mass).

The other substrates for ammonia release include:

- *Urea*　Enzyme involved is urease, e.g. *Bacillus*, *Proteus*, *Micrococcus*, *Sarcina,* etc.

- *Amines*　Substrates are ethylamine, methylamine using the enzyme amino-oxidases produced by *Mycobacterium*, *Pseudomonas*, *Protoaminobacter.*

- *Amides*　*Chlorella*, a green alga utilises amides for producing ammonia. It has the enzyme amidase.

- *Nucleic acids*　The attack is initiated by ribonuclease and deoxyribonuclease.

Factors Influencing Ammonification

Ammonification is more pronounced in well-aerated soils and soils rich in organic matter. In acid soil, there is less ammonia production since decomposition is carried out only by fungi. If the soil is rich in carbohydrate wastes, ammonia formation is low since the microbes prefer to utilise carbohydrates than nitrogenous wastes.

Fate of the Ammonia

Some of the ammonia is volatilised. Since it is positively charged, it is readily bound to the negatively charged clay micelle, thus it does not get washed out of the soil. In the presence of chemoautotrophs and oxygen the ammonia gets readily oxidised and the next step in nitrogen mineralisation, i.e. nitrification process begins.

NITRIFICATION

Nitrification is a two-step process involving ammonium oxidation and nitrite oxidation.

Ammonium Oxidation

This is done by some aerobes called ammonia oxidisers like *Nitrosomonas, Nitrosococcus, Nitrosolobus,* etc.

The ammonia released during ammonification is oxidised to form nitrites. Nitrite is a transient compound and is not usually taken up by plants or microbes. Nitrite ions are toxic and they are readily oxidised to form nitrates which are the assimilable form of nitrogen.

The ammonia oxidising bacteria are chemoautotrophs (obligate) which occur as rods, ellipsoids, spirilla and cocci. They are non-spore forming organisms. Major species of Nitrosomonas are *N. eurospora.* They are gram-negative motile organisms and their optimum growth temperature is 30°C at 7.5–8 pH.

Nitrite Oxidation

The produced nitrite is immediately oxidised to assimilable nitrate and the organisms involved are *Nitrobacter* and *Nitrospira* sp. The organisms are aerobic, gram-negative rods, spherical or spiral shaped. Their optimum temperature is 28°C with a pH requirement of 5.8–8.5.

Factors Affecting Nitrification

Acidity   The nitrifiers are extremely susceptible to acidity even if they produce acids. Thus nitrification proceeds slowly in acid soils.

Oxygen   It is an obligate requirement. Nitrification occurs even in submerged soils (paddy fields) in the upper few centimetres since the diffused oxygen present in waters helps nitrification in such soils.

Moisture   Moisture is needed for nitrification since the nitrifiers cannot tolerate arid conditions.

Temperature   Nitrate production is high during 30°C–35°C, since nitrifiers are mesophiles.

Wastes (organic)   Nitrification depends on C : N ratio. It proceeds faster at a low C : N ratio (when nitrogen-containing organic matter is high).

Fate of Nitrates

They are assimilated by plants and other chemotrophs. It can be easily leached out of soil (through water) since it is negatively charged and is not held firmly by the clay micelle and leads to pollution by eutrophication. It causes infant methaemoglobinemia (blue baby syndrome) due to accumulation of nitrates in tissues (through water, vegetable and fruits) and causes animal methaemoglobinemia.

Agricultural Impact of Nitrification

Since conversion of ammonia to nitrate is very rapid, addition of any ammoniacal fertiliser (for plants) in the soil may increase the competition between plants and ammonia oxidisers (nitrifiers). When nitrate is formed, it is usually wasted since it cannot be retained by the clay micelle (due to its negative charge). Microbes that perform nitrification, use the energy of oxidising ammonia for synthesising organic carbon (carbon fixation). Thus these organisms play dual role, they recycle nitrogen and fix carbon through the energy obtained by recycling nitrogen. This is not economical to the organism, since a lot of energy is required to fix carbon by oxidising ammonia. Hence the nitrifiers are inefficient oxidisers.

DENITRIFICATION

This is the last step in the nitrogen cycle that completes the cycle by replenishing the gaseous nitrogen to the environment. It is the reverse form of nitrification where there is microbial reduction of nitrites and nitrates with the liberation of gaseous nitrogen and nitrous oxide. In denitrification, the nitrogen is lost to the atmosphere and fails to enter the cell structure. The reaction involved is as follows:

$$2HNO_3 \rightarrow 2HNO_2 \rightarrow NO \rightarrow N_2O \rightarrow N_2$$

Microbiology of Denitrification

Arable fields contain an abundance of denitrifying organisms, the population being high in the immediate vicinity of plant roots. Fungi and actinomycetes have not been implicated in nitrogen production. Active bacterial species are *Pseudomonas*, *Bacillus* and *Paracoccus* sp. The denitrifying bacteria are aerobic but in the absence of oxygen, nitrate is used as the electron acceptor. Thus the active species grow aerobically without nitrate or anaerobically in its presence. Most substrates used for aerobic oxidation may be attacked in the absence of oxygen in media containing nitrate.

Several chemoautotrophs are capable of reducing nitrates to molecular nitrogen. *Paracoccus denitrificans*, *Thiobacillus denitrificans*, etc. are some species. Most denitrifiers convert nitrates all the way to nitrogen.

Environmental Factors Affecting Denitrification

- Rate of denitrification is far more slow in soils low in carbon than on land that is rich in organic matter.

- When soil is waterlogged, loss of nitrogen is more when compared to well-drained soils which is due to the immobilisation of inorganic nitrogen.

- Aeration affects the transformation in two ways:

  ♦ Denitrification proceeds only when the oxygen supply is insufficient to satisfy the microbiological demand.

  ♦ At the same time, oxygen is necessary for the formation of nitrites and nitrates which are essential for denitrification. For example, in submerged soils used of rice cultivation, the ammonium is oxidised to nitrate in the oxygen containing surface layer and nitrate is converted to gaseous products of denitrification as it diffuses into the underlying anaerobic zone.

- Many of the bacteria that bring about denitrification are sensitive to high hydrogen ion concentration, hence acid soils contain a sparse denitrifying population.

- The optimum temperature for the reaction is 25°C–60°C.

The enzyme involved in denitrification is nitrate reductase or nitratase. This enzyme is found in *E.coli, Pseudomonas aeruginosa, Micrococcus denitrificans.*

NITROGEN FIXATION

Conversion of gaseous nitrogen into combined form of organic compound by some prokaryotic organisms through biological reactions is called nitrogen fixation. Here nitrogen is immobilized in the organic form within living cells. Broadly, nitrogen fixation can occur in nature in two ways:

1. Non-symbiotic nitrogen fixation.
2. Symbiotic nitrogen fixation.

Non-symbiotic (Free-living) Nitrogen Fixation

This involves diazotrophs which are microorganisms that independently fix atmospheric nitrogen. They are either found in the soil, aquatic habitat or are associated with the rhizosphere (*Azospirillum paspalum*).

Diazotrophs belong to two groups:

Bacteria		Cyanobacteria	
Aerobic	**Anaerobic**	**Heterocystous**	**Non-heterocystous**
Azotobacter	*Clostridium*	*Anabaena*	*Lyngbya*
Beijerinckia	*Desulfovibrio*	*Nostoc*	*Oscillatoria*
Mycobacterium			
Photosynthetic	**Facultative anaerobes**		
Rhodopseudomonas	*Bacillus*		
Chromatium	*Enterobacter*		
Chlorobium	*Klebsiella*		

Special Features of Diazotrophs

The diazotrophs fix atmospheric nitrogen in their specially adapted sites (to suit the acitivity of the enzyme).

Free living bacteria   Bacteria like Azotobacter, Rhodopseudomonas and the like (aerobes) not only fix atmospheric nitrogen within their cells but they also have special adaptation to protect their enzyme (nitrogenase) that is sensitive to oxygen. They produce slime (exopolysaccharide) which retains water and prevents the diffusion of oxygen inside the cell during the process of nitrogen fixation.

Azotobacter evades the harmful effects of oxygen on its nitrogenase by having an exceedingly high rate of respiratory metabolism thus preventing the retention of oxygen inside the cell thus protecting the enzyme complex.

Anaerobes and facultative anaerobes need not have any such special adaptation since their cellular metabolism does not need any oxygen.

In the case of blue-green algae, there are two kinds of diazotrophs, one that possesses special cells called heterocysts and one group that is devoid of it (non-heterocystous).

Many filamentous cyanobacteria contain pale, thick-walled cells (intercalary, lateral or terminal) called heterocysts. This is formed by one of the vegetative cells in the absence of utilisable combined nitrogen (ammonia) since it inhibits heterocyst differentiation and the nitrogen fixing enzyme (nitrogenase). As soon as cyanobacteria growing on ammonia-supplemented media are transferred into nitrogen free medium, both heterocysts and nitrogenase develop simultaneously. The transition of a vegetative cell to a heterocystous cell is a gradual process from an oxygen-evolving cell into an anaerobic cell conducive to nitrogen fixation.

Heterocysts are connected by cytoplasmic bridges to neighbouring photosynthetic cells, thus there is a regulation in flow of molecules between the two types of cells. The vegetative cells supply the necessary photosynthates (for energy) to the heterocysts and they in turn supply fixed nitrogen for growth of vegetative cells.

Heterocysts are the site of nitrogen fixation and they have special mechanisms to protect the nitrogenase complex. They lack the oxygen evolving photosystem II and photosynthetic bile proteins. Walls of heterocysts contain oxygen-binding glycolipids which together with respiratory consumption maintain the anaerobic conditions necessary for nitrogen fixation. In contrast, cells adjacent to heterocysts contain both photosystem I and II, therefore oxygen evolution takes place in these cells. In the case of non-heterocystous nitrogen fixing BGA like *Lyngbya* and *Oscillatoria*, during the nitrogen fixing conditions (in the absence of available nitrogen in the form of ammonia) the filaments are arranged in clumps and nitrogen fixation takes place in the internally organised cells having reduced conditions.

Apart from these organisms, there are certain bacteria like *Azospirillum paspalum* and certain other *Azotobacter* sp. that survive in the microaerophilic conditions associated with the rhizosphere (area or zone surrounding the roots) of paddy plants and fix atmospheric nitrogen in the rhizosphere.

They fix nitrogen only when there is low dissolved oxygen. They make use of the nitrogen gas obtained during denitrification process (on the under surface of stagnant water) and fix it as available nitrogen in the root region of the rice plant.

In summary, nitrogen fixation needs the following:

- *Enzyme* Nitrogenase
- *Energy supply* ATP from photosynthesis, respiration and fermentation
- *Electron acceptors or electron carriers* *Ferredoxin* which is an iron-sulphur protein involved in photosynthesis and pyruvate metabolism and provides the reducing power for conversion of

nitrogen to ammonia. Flavodoxins are produced under limited iron supply by bacteria.

- Protective mechanism for nitrogenase
- Free nitrogen (gaseous form of nitrogen)

Nitrogenase Complex

The following module is a simplified representation of the nitrogenase complex.

Name   Nitrogenase

Class

Oxidoreductases

Acting on reduced ferredoxin as donor

With di-nitrogen as acceptor

Syn-Name

Reduced ferredoxin:dinitrogen oxidoreductase (ATP-hydrolysing)

Reaction

- 3 Reduced ferredoxin + 6 H^+ + N_2 + n ATP = 3 Oxidised ferredoxin + 2 NH_3 + n ADP + n Orthophosphate
- 8 H^+ + 8 e + 16 ATP + N_2 + 16 H_2O = H_2 + 2 NH_3 + 16 ADP + 16 Orthophosphate
- 16 ATP + 16 H_2O = 8 H^+ + 8 e + 16 ADP + 16 Orthophosphate
- 2 H^+ + 2 e + N_2 = Diimine
- 2 H^+ + 2 e + Diimine = Hydrazine
- 2 H^+ + 2 e + Hydrazine = NH_3
- 2 H^+ + 2 e = Hydrogen

Substrate

Reduced ferredoxin

H+

Acetylene

N_2

ATP

Product

      Oxidised ferredoxin

      NH_3

      ADP

      Orthophosphate

Inhibitor   Oxygen

Cofactor

      Iron-sulphur

      Molybdenum

      Homocitrate

Comments

- Possessed by all organisms that can fix nitrogen
- Anaerobic in nature
- Very sensitive to oxygen since O_2 reacts with the iron component of the proteins
- It consists of two brown metallo-proteins (Fig. 42.3).
  - Mo–Fe protein (dinitrogenase) which is the largest unit losing activity at 0°C
  - Fe–protein (dinitrogenase reductase) which is a smaller unit and is less stable than component 1
- Nitrogenase is an equilibrium mixture of Mo–Fe protein and Fe-protein in the ratio 1 : 2
- Nitrogenase needs Mg ions to activate the ATP and requires 12 ATP to reduce 1 mole of nitrogen to ammonia
- Hydrogen is a competitive inhibitor of nitrogenase
- Ammonia represses the synthesis of the enzymes
- Reduces the triple bond compounds in addition to nitrogen like acetylene which is thus useful in quantifying the amount of nitrogen fixed
- Nitrogenase has been purified from many species.
  - Cp type from *Clostridium pasteurianum*
  - Kp type from *Klebsiella pneumoniae*
  - Av type from *Azotobacter vinelandii*. Here Vanadium replaces Mo in the large subunit.

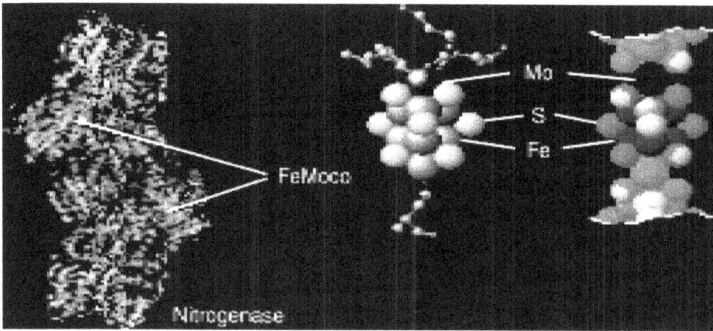

Figure 42.3 *The Nitrogenase enzyme*

Mechanism of Nitrogen Fixation

Chemical nitrogen fixation by Haber Bosch process

- To produce nitrogenous fertilisers
- $N_2 + H_2 \rightarrow NH_3$
- Requires 450°C and 200 atmospheric pressure, hence not cost effective.

Biological nitrogen fixation

- Only possible by limited number of bacterial and archaeal species and some algal species.

    Overall reaction

    $$N_2 + 16ATP + 8H^+ + 8e \rightarrow 2NH_4 + H_2 + 16\ ADP + 16\ Pi$$

    Step 1

    $$N_2 + 2e + 4\ ATP \rightarrow HN=NH\ (diimide)$$

    Step 2

    $$NH_2 - NH_2 + 2e + 4\ ATP \rightarrow NH_2 - NH_2\ (i\ mide)$$

    Step 3

    $$NH_2 - NH_2 + 2e + 4\ ATP \rightarrow 2\ NH_4$$

Apart from this, 4 ATP molecules are wasted by letting out the hydrogen gas during the first step of the reaction. Hence the overall reaction. The efficiency of nitrogen fixation can be increased by the presence of hydrogenase enzymes in an organism (by the possession of 'Hup' genes).

Nitrogen fixation is an energy consuming process, for every mole of nitrogen fixed, about 22 moles of glucose is utilised, i.e. nitrogen fixation is directly proportional to the carbon ratio in the environment, i.e. higher the organic content, higher amount of nitrogen can be fixed since nitrogen fixation also depends on cell mass (rapid multiplication of cells).

Since accumulation of ammonia represses the nitrogenase complex, it is immediately converted to glutamine by glutamine synthase (in the organism). Thus atmospheric gaseous nitrogen gets 'fixed' in the form of ammoniacal nitrogen in the cells of the prokaryotes (diazotrophs).

SULPHUR CYCLE

The sulphur cycle involves compounds in which sulphur exhibits several different valencies. Sulphur cycling includes sulphate reduction to hydrogen sulphide and sulphide oxidation to sulphate. In the presence of oxygen, reduced sulphur compounds are capable of supporting chemolithotrophic microbial metabolism.

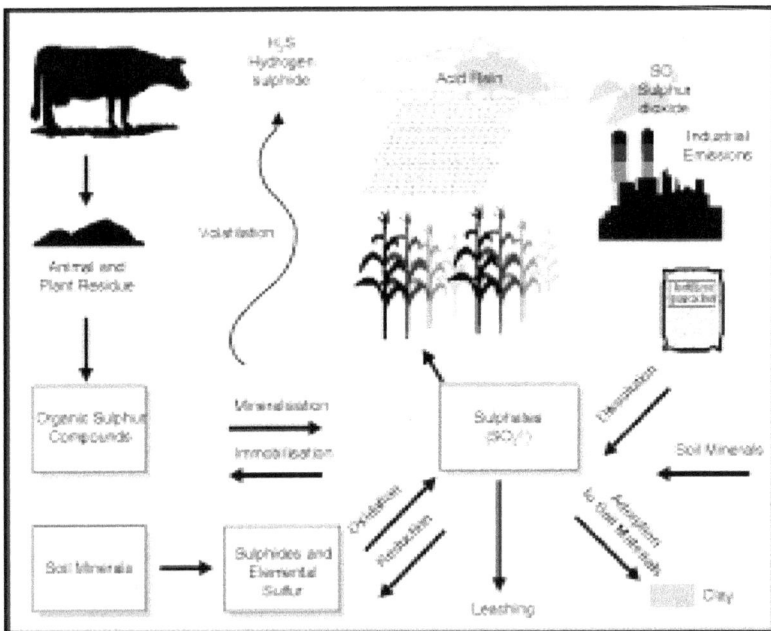

Figure 42.4 *Schematic representation of sulphur cycle*

Sulphur cycle is an oxidation reduction cycle with organic and inorganic pools as shown in Fig. 42.4.

Sources of Sulphur

- *Elemental sulphur*   Sulphur deposits, sulphide ores, post volcanic activities.
- *Inorganic sulphur*   Sulphates, sulphites, thiosulphate, sulphides, sulphur fertilisers of which sulphates are dominant.
- *Organic sulphur*   Plant and animal residues, aminoacids.

Mineralisation of Sulphur

Sulphur is taken up by plants as sulphates and converted to sulphydryl form i.e. reduced form (as in amino acids). Sulphur occurs in amino acids (cystine, methionine), B-vitamins, thiamine, biotin, lipoic acid, tissues and excretory products of animals. On addition to the soil, sulphur contained is mineralised, portion of inorganic sulphur is released and utilised by microflora for cell synthesis and remainder is released into the environment. Aerobic mineralization leads to the formation of sulphates and anaerobic mineralization leads to the formation of hydrogen sulphide and mercaptans. Sulphides are the major inorganic substances released during the decomposition of proteinaceous substrates under anaerobiosis. Mineralization of sulphur in humus is slow but is faster in the presence of oxygen than in its absence. Many bacteria bring about desulphydration of cystine by cystine desulphydrase enzyme liberating hydrogen sulphide. Methionine is resistant to attack. In the oceans, sulphur mineralisation leads to the formation of dimethyl sulphide. Once in the atmosphere, H_2S is oxidised to SO_2 and precipitates as inorganic SO_2 (acid rain).

Sulphur Assimilation

Aerobes prefer sulphur as sulphates while anaerobes assimilate reduced sulphur as H_2S. Heterotrophs generally assimilate organic sulphur (thus they are commonly cultivated in media with sulphur-containing amino acids). Sulphur assimilation is a reductive process. Thus assimilated sulphur is still more reduced. Thirty five percent of assimilated sulphur is combined in C–SH (sulphydryl bonds) of amino acid and glutathione.

Reduction of Sulphur

Once in the inorganic pool, sulphur is subjected to oxidation or reduction.

Some organisms can utilise SO_4 or S_2O_3 (thiosulphate) as the terminal electron acceptors during respiratory metabolism. Microbes involved in sulphur reduction are anaerobic chemoautotrophic bacteria. For example, *Desulfovibrio desulfuricans* is a gram negative curved rod found in soil and sediments. Less than 10% sulphur is assimilated hence these reduction reactions are referred to as dissimilative sulphur reductions. Other bacteria involved in this reaction include *Desulfuromonas acetoxidans*, *Desulfobacter curvatum* and *Desulfovibrio giga*. 'Desulfo' is the precursor given to any organism that reduces sulphur. Sulphur reduction reduces the availability of the sulphur (it is lost as H_2S gas). H_2S is toxic apart from being antimicrobial. Sulphur reduction is retarded by any other electron acceptor like nitrates, ferric or manganous ions. *Desulfovibrio* species usually occurs in anaerobic sediments rich in sulphides and elemental sulphur. It also lives saprophytically with phototrophic green sulphur bacteria and excretes sulphur extracellularly.

$$5H_2 + 2SO_2 \rightarrow 2H_2S + 2H_2O + 2OH^-$$

Thus sulphur reduction also raises the soil pH. In addition to anaerobic sulphur-reducing bacteria, some species of *Bacillus*, *Pseudomonas* and *Saccharomyces* also produce H_2S from SO_4. Many organisms are capable of assimilatory sulphur reduction producing low concentration of H_2S that is immediately incorporated into organic compound (amino acid). Thus difference between dissimilatory and assimilatory sulphur reduction is that in dissimilation sulphur reduction process, H_2S is released into the environment. Sulphur reducing bacteria exhibit a slight preference for ^{32}S over ^{34}S. Biological sulphate reduction is also involved in formation of some elemental sulphur deposits. H_2S released by sulphur reduction may be immediately oxidised to elemental sulphur under anaerobic condition by members of chromatiaceae and chlorobiaceae. Under aerobic condition, H_2S is oxidised to elemental sulphur by *Beggiatoa* and *Thiothrix* group.

Oxidation of Sulphur

In the presence of oxygen, reduced sulphur compounds are capable of supporting chemolithotrophs like *Beggiatoa*, *Thioplaca*, *Thiothrix* and filamentous bacteria capable of oxidation of H_2S.

$$H_2S + \tfrac{1}{2}O_2 \rightarrow S^\circ + H_2$$

Sulphur globules are deposited within the cells. In the absence of H_2S, these globules are slowly oxidised to sulphates. These bacteria are found in the interface of anaerobic environment or the sediment where the partially oxygenated water is in contact with the sediment. Some species of *Thiobacillus* also oxidise the H_2S sine they have low tolerance of acidity and deposit elemental sulphur rather than generating H_2SO_4 by further oxidation. These organisms are called facultative chemolithotrophs. Acidophilic organisms like *T. thiooxidans* that are obligate chemolithotrophs further oxidise elemental sulphur to sulphuric acid. These organisms are acid-tolerant and can withstand up to 2–3 pH and obtain their energy exclusively from oxidation of inorganic sulphur thus aiding in denitrification step of nitrogen cycle. *Thiobacillus denitrificans* utilises nitrates as terminal electron acceptor during oxidation of inorganic sulphur.

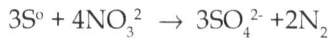

$$3S^\circ + 4NO_3^{2-} \rightarrow 3SO_4^{2-} + 2N_2$$

H_2S is also subjected to phototrophic oxidation in anaerobic environment.

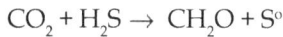

$$CO_2 + H_2S \rightarrow CH_2O + S^\circ$$

Microbial oxidation of reduced sulphur is needed for continued availability of this element in non-toxic form. Sulphur oxidation produces high amount of mineral acids which leads to solubilisation of phosphorus and other mineral nutrients. Typical temperature range for sulphur oxidation is 34–37°C. Mesophiles (*T.ferrooxidans*) are active at 40°C. *Sulfolobus* species are active at 80°C and *S.acidodurans* has an optimum pH of 0.5. It grows in hot acidic sulphate soils. *Beggiatoa* grow in rhizosphere of rice and protect it from H_2S in flooded sediment by its oxidation to elemental sulphur. An important practical implication of sulphur cycle is the anaerobic corrosion of steel and iron structure, destroying pipe set in sulphate containing soils and sediments. The surface of iron reacts with water forming a layer of iron hydroxide and hydrogen.

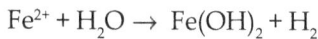

$$Fe^{2+} + H_2O \rightarrow Fe(OH)_2 + H_2$$

Desulfovibrio desulphuricans removes the hydrogen and forms H_2S.

$$H_2 + SO_4^{2-} \rightarrow H_2S + 2OH + 2H_2O$$

The released H_2S attacks iron, forming iron sulphide.

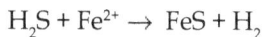

$$H_2S + Fe^{2+} \rightarrow FeS + H_2$$

Thus sulphur cycle is maintained in the respective environments chiefly by the microorganisms involved in the oxidation and reduction of sulphur.

PHOSPHORUS CYCLE

Sources of Phosphates

There is no significant gaseous component of phosphorus and transfer in the air is in particulates or sea spray. All phosphorus exists in phosphate form. It is believed that all original phosphates came from weathering of rocks. Phosphate is found in soil, plants and in microbes in a number of organic and inorganic compounds. It is second only to nitrogen as an inorganic nutrient required by both plants and microbes. The chief source of organic phosphorus compounds entering the soil is the vast quantity of vegetation that undergoes decay. In plants, this element is found in several compounds or in groups of substances as phytin, phospholipids, nucleic acids, co-enzymes, etc. and in animals it is found in bones as well.

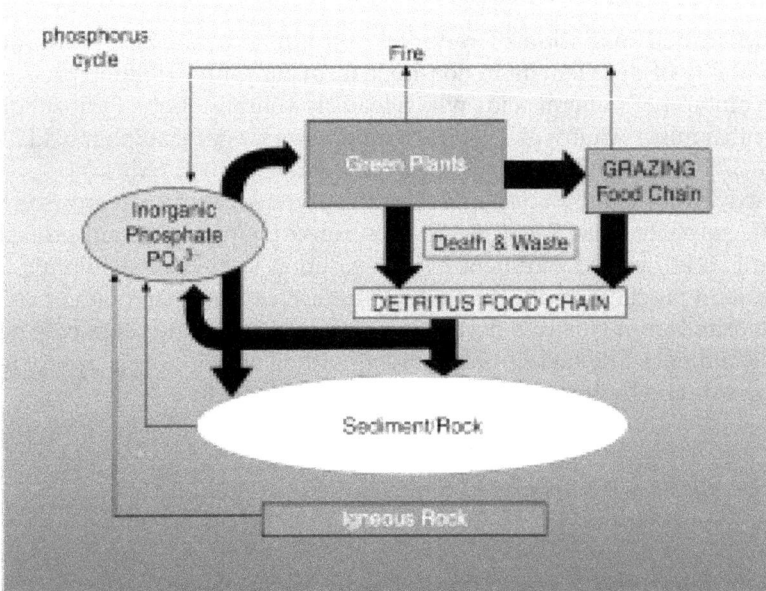

Figure 42.5 *Schematic representation of phosphorus cycle*

Although the total phosphorus content of soils is large, in most soils only a small fraction is available to biota, primarily because of chemical fixation. Through geological time (hundreds of millions of years), phosphorus-containing sediments are buried, uplifted and subject to rock weathering, completing the global cycle. Mining the phosphorus from apatite $(3Ca_3[PO_4]_2 . Ca[FeCl]_2)$, which is primarily biological in origin, is the main source of phosphatic fertiliser (about 14 million metric tons per year).

Phosphorus is easily fixed by chemical reactions with Ca^{2+}, Mg^{2+}, Fe^{2+} etc. P is easily fixed by chemical reactions with Ca^{2+}, Mg^{2+}, Fe^{2+} etc. Both too high and too low pH values can result in P fixation, for example:

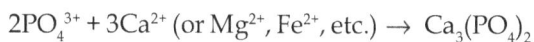

$$2PO_4^{3+} + 3Ca^{2+} \text{ (or } Mg^{2+}, Fe^{2+}, \text{ etc.)} \rightarrow Ca_3(PO_4)_2$$

Microbes play a role in the process of converting organic phosphorus to inorganic phosphate, and in solubilising chemically fixed phosphates (mycorrhizae and rhizosphere activities). They bring about a number of transformations of the element by altering the solubility of inorganic compounds of phosphorus, mineralisation of organic compounds with the release of ionic phosphorus, converting the available inorganic phosphorus into cell components, an immobilisation process and bringing about oxidation-reduction of inorganic phosphorus compounds.

The important steps in the phosphorus cycle are microbial mineralisation and immobilisation as shown in Fig. 42.5.

In contrast to nitrogen, the plant does not reduce phosphate. Phosphate ion enters into organic combination largely unaltered.

Solubilisation of Inorganic Phosphorus

Insoluble inorganic compounds of phosphorus such as rock phosphate (calcium triphosphate) are largely unavailable to plants but many microbes can bring the phosphate into solution. One half of the bacterial isolates are capable of solubilising calcium phosphates, e.g. *Pseudomonas, Mycobacterium, Micrococcus, Bacillus, Penicillium, Aspergillus, Fusarium* etc. Not only do the microbes assimilate the element but they also make a large portion soluble, releasing large quantities in excess to their own nutrient demands. If the insoluble phosphates are suspended in agar medium, the responsible strains are readily detected by the zone of clearing produced around the colony.

The major microbiological means by which insoluble phosphorus compounds are mobilised is by the production of organic acids. The organic or inorganic acids convert calcium triphosphate (rock phosphate) to dibasic and monobasic phosphate with the net result of an enhanced activity of the element to plants. Nitric acid or sulphuric acid produced during the oxidation of nitrogenous materials or inorganic compounds of sulphur react with rock phosphate thereby releasing soluble phosphate. For example, a mixture may be prepared with soil or manure, elemental sulphur and rock phosphate. As the sulphur is oxidised to sulphuric acid by *Thiobacillus*, there is a parallel increase in acidity and a net release of soluble phosphate. Although phosphate solubilisation commonly requires acid production, other mechanisms can account for the same, e.g. ferric phosphate mobilisation in flooded soil. The iron as insoluble ferric phosphates may be reduced, a process leading to the formation of soluble iron with a concomitant release of phosphate into solution. This explains why rice cultivated under water has a lower requirement for phosphorus fertiliser than the crops grown under dry conditions.

The phosphate-dissolving microbes in the vicinity of the roots may enhance phosphate assimilation by higher plants. Enzymes that cleave phosphorus from the organic substrates are called phosphatases. Aspergilli contain acid phosphatases which dephosphorylate and solubilise inorganic and organic phosphorus compounds in phosphatic fertilisers and in the soil. Soil fungi possess another group of enzymes called phytase which helps in the dissolution of organic phosphorus (e.g. Calcium phytate) thereby releasing phosphorus. *Penicillium, Aspergillus, Streptomyces, Bacillus, Pseudomonas* can synthesise this enzyme.

A large number of heterotrophs can develop in media containing nucleotides as the sole source of C, N, and P with the help of ribonuclease and deoxyribonuclease where they cleave the phosphate bond. Bacteria, fungi and actinomycetes are able to use phospholipids as a phosphate source. Mycorrhizal fungi frequently have a dramatic effect on plants whose roots harbour these symbionts. Growth of plants in phosphorus-poor soils is markedly increased if the roots develop mycorrhizal associations in contrast to those not having the fungal association. The symbiotic association frequently allows for extensive phosphate uptake in phosphorus deficient environments as would occur on the addition of fertiliser phosphorus.

Immobilisation

Microbial growth requires the presence of available forms of phosphorus. The assimilation of phosphorus into microbial nucleic acids, phospholipids leads to the accumulation of non-utilisable forms of the element. Hence during the decomposition of organic matter added to soil, the increase in microbial abundance puts a great demand on the phosphate supply. Phosphorus, like nitrogen, is therefore both mineralised and immobilised. The process that predominates is governed by the percentage of phosphorus in the plant residue undergoing decay and the nutrient requirements of the population which exceed the concentration of that required for microbial nutrition. The excess appears as soluble inorganic phosphate, and if inadequate for the microflora the net effect is immobilisation. As long as there is decay and microbial cell synthesis, both mineralisation and immobilisation will take place.

Soil-Based View of Phosphorus Cycle

Initially, phosphate weathers from rocks. The small losses in a terrestrial system caused by leaching through the action of rain are balanced in the gains from weathering rocks. In soil, phosphate is absorbed on clay surfaces and organic matter particles and becomes incorporated (immobilised). Plants dissolve ionised forms of phosphate. Herbivores obtain phosphorus by eating plants, and carnivores by eating herbivores. Herbivores and carnivores excrete phosphorus as a waste product in urine and faeces. Phosphorus is released back to the soil when plant or animal matter decomposes and the cycle repeats.

Global View of Phosphorus Cycle

The phosphorus cycle occurs when phosphorus moves from land to sediments in the seas and then back to land again. The main storage for phosphorus is in the earth's crust. On land phosphorus is usually found in the form of phosphates. By the process of weathering and erosion phosphates enter rivers and streams that transport them to the ocean. Once in the ocean the phosphorus accumulates on continental shelves in the form of insoluble deposits. After millions of years, the crustal plates rise from the sea floor and expose the phosphates on land. After more time, weathering will release them from rock and the cycle's geochemical phase begins again.

Ecosystem View of Phosphorus Cycle

The ecosystem phase of the phosphorus cycle moves faster than the sediment phase. All organisms require phosphorus for synthesising phospholipids, NADPH, ATP, nucleic acids, and other compounds. Plants absorb phosphorus very quickly, and then herbivores get phosphorus by eating plants. Then carnivores get phosphorus by eating herbivores. Eventually both of these organisms will excrete phosphorus as a waste. This decomposition will release phosphorus into the soil. Plants absorb the phosphorus from the soil and they recycle it within the ecosystem.

Factors affecting solubilisation of insoluble phosphates

Physical factors

Temperature   Optimum temperature for *Pseudomonas striata* is 20–30°C. For fungi the temperature is 30°C. Solubilisation is drastically reduced at high temperatures.

Aeration   There is significant increase in rock phosphate solubilisation in shake cultures than in static cultures of *P.striata* and *Aspergillus awamori*. This explains the importance of aeration.

Incubation period   Under cultural conditions, phosphorus solubilisation takes up to 10–15 days. Generally solubilisation increases with incubation period.

Rock phosphate concentration   If concentration of phosphorus dioxide is higher than 100 mg/100 ml, there is improvement in solubilisation by bacteria and fungi. Lower the quantity of phosphate in the medium, greater is the solubilisation by Aspergillus awamori.

Particle size of rock phosphates   Maximum solubilisation occurs when the rock phosphate particle size is between 30–59 and 60–99 mesh. Efficiency of microbial solubilisation is reduced in the presence of finer rock phosphate particles of 150–250 mesh as compared to coarser particles.

Chemical factors

Carbon source   Different carbon sources in the form of monosaccharides, disaccharides and alcohols have been found to affect phosphorus solubilisation. Glucose or xylose is the best source of energy for fungi in liquid medium. Glucose, sucrose and galactose are good sources of energy and carbon for *P.striata*. By increasing the concentration of glucose as a carbon source from 1–3% in liquid medium, amount of phosphorus solubilised was enhanced.

Nitrogen source   *P. striata* utilises ammonium sulphate, asparagine, urea, ammonium nitrate and calcium nitrate but it cannot use sodium nitrate. *Schwanniomyces occidentalis* (yeast) can only use ammoniacal form of nitrogen and not nitrate form.

pH   Optimum pH for bacterial solubilisation is 6. For fungi and yeast it is 5-6. Fungal activity was greatly reduced at 7–8.

Application in Agriculture

Phosphocompost is a phosphorus source in neutral and alkaline soils. Phosphocomposting involves the incorporation of rock phosphorus ensuring composting of organic wastes in the ratio of 2.5–3 and 7–7.5. During composting, a lot of organic acids and humic substances are produced as a result of microbial activity which bring about solubilisation and account for 75–90% of the phosphorus solubilised and the phosphorus enriched compost is called phosphocompost.

Interaction of Phosphate-solubilising Microbes with VAM

Common genera of VAM like *Glomus*, *Gigaspora*, *Acaulospora*, *Sclerocystis* and *Entophosphora* grow and increase phosphate uptake from the soil. They have the ability to improve the utilisation of unprocessed phosphatic urea or the rock phosphate which contain sparingly soluble phosphorus. Rock phosphate is the economic source of phosphorus particularly in acid soils and dual inoculation of phosphate-soluble bacteria *Bacillus circulans* and the VAM fungus *Glomus* sp. produces a synergistic effect by bringing about simultaneous solubilisation of insoluble phosphate and mobilisation of the same. Application of high amounts of soluble phosphates lowers mycorrhizal level. This problem is overcome by applying high amounts of insoluble low-grade rock phosphate along with a dual inoculum of VAM fungus and a phosphate solubilising microbe.

IRON CYCLE

The cycling of iron consists largely of oxidation–reduction reactions that reduce ferric iron to ferrous iron and oxidise ferrous iron to ferric iron. Ferric iron precipitates in alkaline environments as ferric hydroxide. Ferric iron may be reduced under anaerobic conditions to the more soluble ferrous form. Under some anaerobic conditions, however, sufficient H_2S may be evolved to precipitate iron as ferrous sulphide. Flooding of soil, which creates anaerobic conditions, favours the accumulation of ferrous iron. In aerobic habitats such as well drained soil, most of the iron exists in the ferric state. Microbial growth often is limited by the availability of iron. Various bacteria produce siderophores which bind iron and facilitate its cellular uptake. Some chemolithotrophs oxidise iron to generate cellular energy. These iron-oxidising bacteria can lead to substantial iron deposits.

Iron is the fourth most abundant element in the earth's crust. Microbial metal transformations are essential for the production of metallic ores. They are important in extracting metals from low-grade ores.

Microbial iron transformations include:

- Iron scavenging and uptake
- Iron oxidation and precipitation
- Iron reduction and solubilisation

In aerobic environment, microbial iron oxidation dominates in acidic condition and chelation dominates in neutral environment. In the anaerobic environment, iron cycle is dominated by iron reduction and precipitation of iron sulphides.

Iron exists as metallic ion (ferric and ferrous ion). Metallic iron spontaneously oxidises in acidic condition:

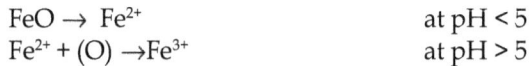

$$FeO \rightarrow Fe^{2+} \qquad\qquad\qquad \text{at pH} < 5$$
$$Fe^{2+} + (O) \rightarrow Fe^{3+} \qquad\qquad \text{at pH} > 5$$

spontaneous reactions are slow

Under aerated conditions, bacteria obtain energy from oxidation of ferrous ions (Fe^{2+}) (e.g. *Thiobacillus ferrooxidans* and *Sulfolobus acidocaldarius*) are iron oxidising bacteria.

$$2Fe^{2+} + \tfrac{1}{2} O_2 + 2H^+ \rightarrow 2Fe^{3+} + H_2O$$

T. ferrooxidans is a motile chemolithotrophic bacillus, which grows at pH 2–4. It oxidises the iron in ferrous sulphate to ferric sulphate.

Benefits of Iron Oxidation

Iron oxidation allows microbes to gain energy. The insoluble ferric ion that is produced, physically protects the microorganisms. Many bacteria like *Leptothrix*, *Seliberia Sphaerotilus* oxidise ferrous ion without gaining any energy. Ferric ion gets precipitated around filamentous sheath forming bacteria. Other iron oxidisers are *Ferrobacillus sulfoxidans*, *F. ferrooxidans*.

Iron Reduction

Many bacteria and fungi reduce ferric ion, e.g. *Alternaria*, *Bacillus*, *Clostridium*, *Fusarium*, *Klebsiella*, etc. Iron is a good electron acceptor. Oxygen and nitrates suppress iron reduction. Iron reduction occurs under anaerobic conditions resulting in the accumulation of ferrous ion. Although many microbes can reduce small amounts of iron during their metabolism, most iron reduction is carried out by specialised iron-respiring microbes such as *Geobacter metallireducens, Geobacter sulphurreducens, Ferribacterium imneticum* and *Shewanela putrefaciens*, which can obtain energy for growth on organic matter using ferric iron as an oxidant.

Gleying   It is a term applied to a condition in which iron in soil is reduced and has a grayish green colour. It reflects microbial use of iron as an electron acceptor for oxidising carbon substrates.

Iron Availability and Assimilation

Ferric ion is highly insoluble but it can be solubilised by acidification and combination with organic material. Ferric ion and organic acid in forest soil becomes more soluble and percolates through the soil profile. Ferric ion precipitates in the B horizon and forms a distinct layer.

Ferric ion is unavailable when compared to ferrous ion. Iron solubility is poor in alkaline soils and leads to chlorosis (leaf yellowing due to iron deficiency shown through non-synthesis of chlorophyll).

Iron is removed from the environment by organic chelating compounds like citrate, oxalate, humic acids and tannins, siderophores, ferritin (iron storage compounds). In additon to these relatively simple reduction to ferrous ion, some magnetotactic bacteria such as *Aquasirillum magnetotacticum* transform extracellular iron to the mixed valence iron oxide mineral magnetite and construct intracellular magnetic compasses. Furthermore, dissimilatory iron reducing bacteria accumulate magnetite as an extracellular product.

Review Questions

1. Mention the sources of organic and inorganic carbon.
2. Discuss in detail the carbon cycle in the environment.
3. Describe the nitrogen cycle with an illustration.
4. Discuss the microbiology of denitrification.
5. Briefy discuss the various factors affecting each step in the nitrogen cycle.
6. Discuss about diazotrophs.

7. Comment on nitrogen fixation as an integral part of nitrogen cycle.

8. Discuss in detail about nitrogenase complex.

9. Explain the sulphur cycle in detail.

10. Discuss the iron cycle in detail.

11. What is phosphorus solubilisation? Explain.

12. What is the effect of VAM in phosphate solubilisation?

43

Nitrogen Fixation (Symbiotic)

Nitrogen is a constituent of amino acids, nucleic acids, amino sugars and their polymers. But availability of combined nitrogen is an important limiting factor for primary production in an ecosystem. A number of free-living microorganisms can assimilate molecular nitrogen but no higher plant or animal has the needed enzyme to catalyse the reaction. Symbiosis can overcome this problem where one of the important effects is to fix atmospheric nitrogen.

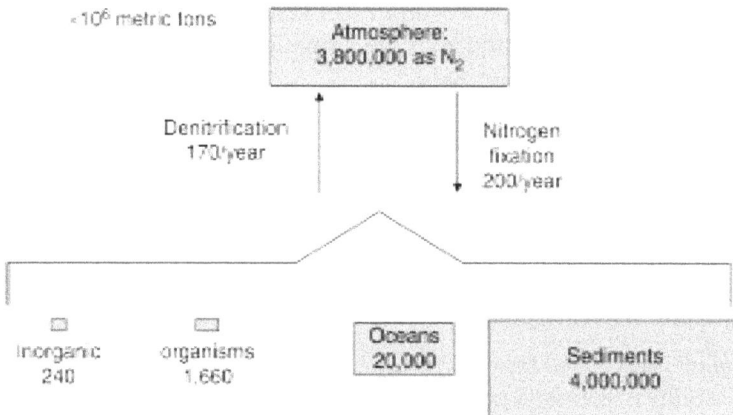

Figure 43.1 *Data on the turnover of nitrogen fixation*

RHIZOBIUM

A classical example of nitrogen fixation is the symbiosis between leguminous plants and bacteria of the genus *Rhizobium*. The seat of symbiosis is within the nodules that appear on the plant roots. In the symbiotic association between *Rhizobium* and leguminous plant, two organisms interact in such a way as to influence and coordinate the expression of both prokaryotic and eukaryotic gene.

The communications that occur between the plant and the rhizobia during nodule formation and maintenance constitutes a novel opportunity to study signal transduction in a plant system. The expression of 'nodulation' genes in the bacteria is activated by signals from plant roots and as a result the bacteria synthesise signals that induce a nodule meristem and enable the bacteria to enter this meristem via a plant-made infection thread. The chemical signals synthesised by the bacteria are based on a modified amino acid (homoserine lactone) carrying a variable acyl chain substituent, and are called acyl homoserine lactones (AHLs). By detecting and reacting to these chemicals, individual cells can sense how many cells surround them, and whether there are enough bacteria, i.e. a quorum, to initiate the change towards acting in a multicellular fashion. This is known as 'quorum sensing'.

Beijerinck was the first to isolate and cultivate the bacterium responsible for nodulation and he named it *Bacillus radicicola* which is now placed in the genus *Rhizobium*. *Rhizobium* is found as a free-living organism in the soil but does not fix atmospheric nitrogen in that state. Only after its association with a leguminous plant and after the formation of root nodules does it fix atmospheric nitrogen in the nodules.

Source of Inoculum in the Soil

When a nodule becomes senescent after a period of nitrogen fixation, decay of tissue sets in liberating motile forms of *Rhizobium* in the soil which normally serves as a source of inoculum for the succeeding crop.

Classification of Rhizobium

Rhizobium is classified based on cross inoculation grouping which is based on the ability of a limited genera or species capable of nodulating groups of legumes related to one another. All rhizobia that could form nodules on roots of certain legume have been collectively taken as a species, e.g. *R. leguminosarum* nodulates the pea group consisting of *Pisum*, *Vicia* and

Lens. R. phaseoli infects the bean group and *Rhizobium trifolii* infects the clover group. Recent version of rhizobial classification has decided that *Rhizobium* will have three recognised species *R. leguminosarum*, *R. meliloti* and *R. loti*.

Cultural Characteristics of Rhizobium

Rhizobium can live on relatively simple synthetic medium but it is incapable of fixing nitrogen. It can fix nitrogen only in the root nodules of leguminous plants (Fig. 43.2). Fast growing *Rhizobium* like *R. trifolii*, *R. leguminosarum* can utilise all sources of carbohydrates but slow growing *Rhizobium* are specific in nutrients and they need simple sugars like xylose, mannitol, arabinose for their growth. *Rhizobium* cells are small to medium sized gram negative bacilli. They are non-spore formers. They are motile when young. Cells contain characteristic granules of B-hydroxybutyrate granules which stain with sudan black and appear as highly refractile bodies. Most strains produce gum (extracellular polysaccharide slime).

Figure 43.2 *Rhizobia within the nodule*

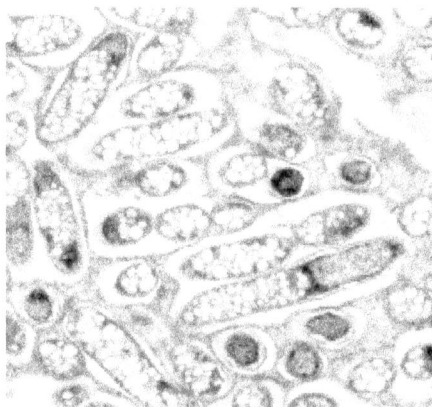

Ecology

Nodule bacteria occur in soil and in the root region of legumes as well as non-legumes. In the absence of legumes soil population of rhizobia decline. Rhizobia present in the soil may be antagonised by other bacteria, fungi or phages. *Rhizobium* is very acid sensitive. Soil acidity may eliminate *Rhizobium*. Generally *Rhizobium* can tolerate high soil temperature.

Infection

A functional root nodule is a site of symbiotic nitrogen fixation. It is developed by cooperative interactions between the legume host and *Rhizobium*. Thus an effectively nodulate legume is fully autotrophic for nitrogen, a limiting factor in plant nutrition. Specific group of legumes are receptive to a particular species of *Rhizobium*. Initiation of nodule development requires the successful entry of *Rhizobium* into the host root cell which generally takes place through the root hair. The first visible plant response to *Rhizobium* is the curling or contorted growth of the hair at the portion of infection followed by the development of a tube-like structure called the infection thread (IT). The subsequent growth of the IT and pattern of infection of root cortical cells is determined by the host. In legumes such as *Arachis* (groundnut), where an IT is not observed, it appears that infection proceeds through intracellular spaces.

Rhizobium enters into the root hair either through intracellular gaps or through the formation of IT.

Host—Rhizobium Specificity

Recognition of a specific host by *Rhizobium* is a prelude to infections. Specific plant proteins (phytohaemaglutinins) called lectins may be involved in this process. For example, lectin from *Phaseolus* can bind *R. phaseolii* at root sites suitable for infection. These molecules provide a host-rhizobium specificity. The biochemical receptor for lectin binding appears to be an 'O' antigen moiety of an LPS of *Rhizobium*. The lectins bind specifically to the capsular form of *Rhizobium*. The strains of *Rhizobium* that do not bind lectins are incapable of nodulation. Usually in a population of *Rhizobium*, less than 1% of the bacteria appear to differentiate into a capsular form, thus only a very small percentage of cells may actively participate in the infection process. According to some, a phenomenon of chemotaxis is involved in attracting specific *Rhizobium* and is mediated by a glycoprotein called chemotactin. It has been suggested that this protein may provide some specificity for attractions for *Rhizobium* by a host involvement of cellulolytic and pectic enzyme in infection.

Rhizobium does not appear to destroy the wall of the root hair completely at the period of attachment. Instead, the cell wall of the hosts invaginated at the site of infection and due to the deposition of new cell wall material an IT is formed through which Rhizobium enters the host cell. Obviously, an alteration in the cell wall structure is a prerequisite for such an invagination.

It was proved that activities of cellulase and pectinase enzymes increase during early development of nodule and that these activities are compartmentalised, i.e. pectinase is primarily present in the bacteria and cellulase in the host. The bacterial pectinases are primarily responsible for initiation of infection or the formation of *shepherd's crook*. Eventual dissolutions of the IT wall however takes place with the help of cellulase produced by the host. Thus a cooperative action of both host and *Rhizobium* are involved in the initiation and development of infection during root nodule formation. Since cellulase induction is under phytohormone control and *Rhizobium* can produce these hormones, the production of cellulase may be indirectly controlled by *Rhizobium*. This appears to be an intense interaction between the nucleus of root hair cell and IT originating at the tip of the curled portion of the root hair. The nucleus guides the path of IT in the root hair.

Release of Bacteria from IT

A successful infection results in the release of bacteria from the IT into a membrane vesicle derived from the host polysaccharide matrix. Upon the entry of the IT into the cortical cells of the root, the thread branches and thus traverses intracellularly. The contents of an IT (bacteria) are liberated into a tetraploid cell which is then induced to intense meristematic activity which is again controlled by phytohormones. Sooner or later, well-differentiated areas are demarcated showing a diploid nodule cortex with a central tetraploid bacteroid zone having vascular connection with the parent root system.

Intracellular Compartmentalisation of Rhizobium

Following dissolution of the IT wall, *Rhizobium* may be considered released into the host cell cytoplasm though they remain bound in membrane envelope. Thus physiologically they are extracellular. In an effective root nodule, no bacteria have been observed to be free inside the host cell. Release of bacteria from the IT is followed by an increase in plasma membrane biosyntheses which keeps pace with the bacterial proliferation. As a result of an increased population of bacteria and a high demand for oxygen, the intracellular environment of the host cell becomes hypoxic. In this environment, bacteria differentiate into *bacteroids*. Once the rhizobia are released, they lose their motility and reproducing abilities. They round off and may exist in various shapes like 'Y', 'L' etc. Now they are called bacteroids. Depending on the legume, bacteroids are surrounded by membrane envelopes whose origin is hypothetical.

- They are formed *de novo* after the release of the bacteria from the IT.

- They are extensions of the endoplasmic reticulum of host cells.

- They are derived from plasma membrane by a process of phagocytosis.

The number of bacteroid in any envelope is determined by the host. Large number of bacteroid per membrane envelope may be due to the fact that the membrane does not grow fast enough than the bacteria. The root nodule can be subdivided into specific zones. The outer periderm consists of loosely packed cells and lenticels through which gases diffuse into the outer cortical layer. These gases include the N_2 required for N_2 fixation, as well as O_2 required for plant and bacterial respiration. The cells of the outer cortex are also loosely packed, and contain large air spaces which offer little resistance to the diffusion of gases. However, in the inner cortex, the cells are smaller and much closer together, and here the gases may have to diffuse through the contents of the cortical cells of ammonia to reach the central zone of the nodule, since open intercellular spaces are infrequent. This part of the nodule is thought to act as a barrier to gas diffusion.

The central zone of the nodule contains plant cells that are infected with thousands of bacteria. In their symbiotic form, these bacteria are called bacteroids, and a typical soybean root may provide a home for 2,000,000,000 of these bacteroids. The bacteroids are responsible for reducing the N_2 gas that diffuses into the central zone to ammonia, which can then be assimilated by the plant cells. More important is the exchange of materials that must occur between the plant and the bacteroids to allow for the fixation of N_2 and the assimilation. A carbon source from the plant is supplied via the phloem to root nodules housing the N_2 fixing bacteria. The carbon source is partially metabolised by the plant and the resulting carbon compounds are imported into the bacteria. The bacteria contain the enzyme nitrogenase, which is responsible for catalyzing the reduction of N_2 gas to ammonia. The carbon compounds entering the bacteria are metabolized to produce the ATP and reductant required in the nitrogenase reaction. The ammonia produced in the reaction is transported to the plant where it is assimilated into organic nitrogenous compounds. These are exported in the xylem to the rest of the plant where they are used in the synthesis of amino acids, nucleic acids and other nitrogen-containing compounds.

ROLE OF PHYTOHORMONES IN NODULATION:

Presence of indole acetic acid (IAA) and kinetin are confirmed in the root nodules where several species of *Rhizobium* are able to convert tryptophan to IAA. Tryptophan produced in the legume roots is converted to IAA by the bacterium and thus auxins are synthesised symbiotically. As the IT progresses through the root hair into the cortical tissue, cells divide and proliferation activity starts. This stimulus is due to auxin and cytokinin which causes endoreplication in some cortical cells (the reason for the presence of tetraploid cells in the cortex). Auxins are also known to induce hydrolytic enzyme activity such as cellulase.

Function of the Nodule

- Bacteroids are the site of nitrogen fixation. There are two main requirements for biological reduction of atmospheric nitrogen:

  - Continuous availability of carbohydrate for energy and ATP

  - Protection of *Rhizobium* nitrogenase from oxygen.

- *Legume   Rhizobium* association produces an environment in the nodules where these requirements are met by the host. The host provides the photosynthates (organic acids, amino acids and carbohydrates) to *Rhizobium* which metabolizes these compounds to generate reductants and ATP for the nitrogenase reaction.

- The nodules continue to fix nitrogen at a reduced rate in the dark as well as when they become detached from the root for a long time indicating that they have higher reserves of carbohydrates in the form of polyhydroxybutyrate granules.

- There is a high demand for oxygen in nitrogen fixing tissue and in order to maintain a continuous flow of oxygen, host cells produce an oxygen binding protein, leghaemoglobin in the infected cells.

- Leghaemoglobin is a red pigmented protein similar to haemoglobin. 'Leg' indicating presence in legume. It is a haemprotein having a haem moiety attached to a peptide chain which represents the globin part of the molecule.

- The haem part is produced by the bacteria and globin part is produced by the plant. Leghaemoglobin is confined to the central nodule cells containing the bacteroids.

Measurement of Nitrogenase Activity

Nitrogenase is a promiscuous enzyme that is capable of reducing a wide range of substrates. The traditional method of assaying nitrogenase activity is by the acetylene reduction assay, in which nodulated roots are supplied with 10% acetylene. This gas is reduced by nitrogenase to ethylene, and the amount of ethylene produced in a measured time period is measured by gas chromatography. This assay is dangerous to perform, since 10% acetylene is explosive, and it uses a non-physiological substrate for the nitrogenase reaction. A better assay is to measure the H_2 that is produced as an obligate by-product of the N_2 fixation reaction:

$$N_2 + 8H^+ + 8e^- + 16ATP \rightarrow 2NH_3 + H_2 + 16ADP + 16Pi$$

Many *Rhizobium* and *Bradyrhizobium* species possess an enzyme called uptake hydrogenase (HUP) which re-oxidises the H_2 produced in the nitrogenase reaction. However, many *Rhizobium* and *Bradyrhizobium* species lack this enzyme, and the H_2 produced within the nodule diffuses out into the soil. In the following experiments, the soybean roots have been inoculated with a bacterial species (*B. japonicum* USDA 16) that lacks HUP and therefore evolves H_2. The rate of H_2 evolution from the nodules provides a measurement of nitrogenase activity.

The great advantage of measuring nitrogenase activity by H_2 evolution is that measurements can be made with minimal disturbance of the plant tissue. Nitrogenase activity is measured by a technique called Gas Exchange Analysis in which a gas mixture (such as air) lacking H_2 is passed across a nodulated root system, in a sealed cuvette, at a known flow rate. The H_2 content of the gas after it has passed across the root is measured to determine the rate of H_2 evolution from the nodules. This technique requires the use of a Gas Exchange System as described in the following section on Set-Up Procedures.

Measurement of H_2 production from nodules provides only a measurement of Apparent Nitrogenase Activity (ANA). This is because of the 8 electrons passing through nitrogenase in the reaction shown above, a minimum of 2 of these are used to produce H_2, while the others are used in the reduction of N_2 to ammonia. H_2 production rate provides only a measurement of that proportion of the enzyme activity that is being used to reduce protons, and while this must represent at least 25% of the total enzyme activity, the allocation of electrons between N_2 and H^+ reduction is variable. To measure Total Nitrogenase Activity (TNA), it is necessary to replace the

N_2 in the atmosphere surrounding the nodule with an inert gas such as Ar. When this is done, all the electrons that were previously allocated to N_2 reduction are used for H^+ reduction and the rate of H_2 evolution from the nodule increases rapidly. The maximum rate of H_2 evolution from the nodule after a switch from a $N_2:O_2$ atmosphere to an $Ar:O_2$ atmosphere represents TNA. To determine the rate of N_2 fixation that was occurring under initial conditions the following equation is used:

$$N_2 \text{ fixation rate} = (TNA - ANA) / 3$$

A denominator of three is used because reduction of N_2 to NH_3 requires 3 electron pairs, whereas reduction of H^+ to H_2 requires only one electron pair.

As stated above, in air at least 25% of total electron flux through nitrogenase is used for proton reduction but this value is not constant, and often more than 25% of electron flux through nitrogenase is 'wasted' in H_2 production. To determine the relative allocation of electrons between H^+ and N_2, the Electron Allocation Coefficient of nitrogenase (EAC) may be calculated thus.

Regulation of Nitrogenase Activity

The nitrogenase reaction requires a great deal of energy, consuming at least 16 ATP and 4 pairs of electrons for every molecule of N_2 reduced to ammonia. This energy is derived from carbohydrates provided by the plant, and nitrogenase activity therefore represents a drain on the plant's resources. To ensure that nitrogenase functions are beneficial rather than detrimental to the plant, regulatory mechanisms have evolved to reduce, or to halt, nitrogenase activity when the plant is supplied with an adequate source of combined nitrogen. Since carbohydrate supply to the nodule is essential for the continuance of nitrogenase activity, it is no surprise that any factor which limits phloem sap supply to the nodule also causes nitrogenase inhibition. However, the mechanisms by which nitrogenase is inhibited by phloem sap deprivation, or by combined nitrogen, are complex, and require an understanding of the regulation of respiratory processes in the nodule as outlined below.

Nitrogenase Activity and Oxygen

The nitrogenase enzyme is peculiar in that, to function it requires ATP derived from oxidative phosphorylation, although it is extremely O_2-labile. Therefore, the nodule must have mechanisms for delivering a high flux of O_2 to support respiration in the infected cells, while maintaining the free O_2 concentration in these cells at an extremely low level. It has been estimated that the O_2 concentration in the infected cells of the nodule may be as low as 30 nM, compared to an O_2 concentration of 250, 000 nM in the cells of the nodule outer cortex. This low O_2 concentration in the infected cells is maintained by a barrier to O_2 diffusion in the nodule inner cortex which severely limits the diffusion rate of O_2 to the central zone. Also, the high respiration rate of bacteroid respiration ensures that the O_2 diffusing into the central zone is consumed swiftly and does not accumulate. In fact, the O_2 concentration in the infected cells is so low that O_2 cannot diffuse fast enough by itself to support the demands of bacteroid respiration. To alleviate this limitation, the infected cells contain a red pigmented protein called leghaemoglobin which acts as an O_2 carrier, and facilitates the diffusion of O_2 to the bacteroids. In both its structure and its function, leghaemoglobin is very similar to myoglobin in human blood, and it is an example of how plants and animals have evolved similar mechanisms to cope with similar problems. The leghaemoglobin in nodules is easily visible, and its red color clearly delineates the central zone of the nodule from other nodule tissues.

Regulation of the O_2 Diffusion Barrier

It has been shown that the barrier to O_2 diffusion in the legume nodule does not have a fixed resistance, but can vary the degree to which it limits O_2 diffusion into the central zone. If the nodulated root is placed in a low O_2 environment, the diffusion barrier relaxes its resistance to maintain an adequate flux of O_2 to the infected cells. Conversely, when the nodulated root is placed in a high O_2 environment, the diffusion barrier increases its resistance to prevent too much O_2 entering the infected cells and causing nitrogenase inhibition. Although low O_2 environments are common in nature (such as in waterlogged soils), environments containing supra-ambient levels of O_2 are extremely rare. However, the ability of the nodule to respond to high levels of O_2 in the infected cells is very important, since any process which limits the respiration rate of these cells will reduce the rate at which O_2 is consumed and cause O_2 to accumulate. Therefore, under these circumstances the nodule increases its resistance to O_2 diffusion to reduce the flux of O_2 to the central zone. It is this increase in diffusion barrier resistance that is mostly responsible for the inhibition of nitrogenase activity that occurs when nodules are supplied with combined nitrogen, or are deprived of phloem sap.

When combined nitrogen is supplied to the nodulated root, continued nitrogenase activity would represent a waste of the plant's carbohydrate resources. Therefore, a signal (as yet unidentified) is given to the nodule to increase its resistance to O_2 diffusion. This causes a reduction in the rate of infected cell respiration, and insufficient ATP and reductant are generated for the nitrogenase reaction to continue at a high rate. When the supply of phloem sap to the nodule is prevented (e.g. by severing the plant shoot (detopping), or removing its leaves (defoliation), or by picking the nodules off the plant), the reduced amount of carbohydrate entering the nodule could eventually lead to a reduction in infected cell respiration and a consequent increase in infected cell O_2 concentration. Therefore, a signal (again, as yet unidentified) is given to the nodule to increase its resistance to O_2 diffusion. This has the combined effect of limiting O_2 entry into the central zone (thereby protecting nitrogenase from inactivation) and of reducing respiration rate (thereby conserving carbohydrates that may eventually become scarce).

Review Questions

1. Discuss the cultural and ecological characteristics of Rhizobium.
2. Describe in detail the process of nodule formation.
3. Discuss at length the intracellular compartmentalisation of Rhizobium.
4. Briefly explain the function of nodules.
5. Explain the mechanism of regulation of nitrogenase activity.

PART VIII

WASTE WATER MICROBIOLOGY

Sewage

Microbiology of Sewage

Biochemical Oxygen Demand

Chemical Oxygen Demand

Sewage Treatment

Applications of Waste Treatment

44

Sewage

Sewage or wastewater is the waterborne human, domestic and farm wastes. It may include industrial effluents, subsoil or surface waters, human wastes include faecal material. Domestic wastes include food wastes and wash water. Industrial waterborne wastes are acids, oils, greases and animal and vegetable matter discharged by factories. The composition of sewage varies depending upon the source of waste water. This also causes variation in the microbial flora of sewage. Almost all groups of microbes, algae, fungi, protozoa, bacteria and viruses are present.

The waste water discharged through the drainage system has to be properly disposed. They cannot be simply disposed off into water bodies or landscapes because of the oxygen demand they exert and also due to the presence of pathogenic microbes in them. So before disposal, waste water has to be properly treated.

Rain water and ground water are very pure. But human activities like discharge of noxious substances including biocides, oil and sewage into water bodies spoil the aquatic environments to the maximum. Increased growth of microorganisms also contribute to the deleterious effects on aquatic environments by affecting the ecological balance with complicated consequences, as in eutrophication. The heterotrophic activity of microorganisms in water depletes the dissolved oxygen content of water which has a negative influence on the self- purification of water. This leads to what is known as biological oxygen demand or BOD.

Liquid wastes are produced by everyday human activities (domestic sewage) and by various agricultural and industrial operations. Following drainage patterns or sewers, liquid waste discharges enter natural bodies of surface water such as rivers, lakes and oceans. At much slower rates, they may also percolate to the ground water table, especially if it is high or if fissures are present in the unsaturated soil layer.

Natural waters have an inherent self-purification capacity. Organic nutrients are utilised and mineralised by heterotrophic aquatic microorganisms. Allochthonous populations of enteric and other pathogens are reduced in number and eventually eliminated by competition and predation pressures exerted by the autochthonous aquatic populations.

Waste treatment, protection of drinking water sources, and disinfection of drinking water and sewage, gradually introduced in the early years of this century, largely eliminated the spread of waterborne pathogens. Waste treatment, protection of drinking water sources, and disinfection of drinking water and sewage, gradually introduced in the early years of this century, largely eliminated the spread of waterborne epidemics in developed countries.

There are various options available to convert solid waste to energy. Mainly, the following types of technologies are available: (1) sanitary landfill, (2) incineration, (3) gasification, (4) anaerobic digestion, and (5) other types.

Sanitary landfill is the scientific dumping of municipal solid waste due to which the maturity of the waste material is achieved faster and hence gas collection starts even during the landfill procedure. Incineration technology is the controlled combustion of waste with the recovery of heat, to produce steam that in turn produces power through steam turbines. About 75% of weight reduction and 90% of volume reduction is achieved through burning. A gasification technology involves pyrolysis under limited air in the first stage, followed by higher temperature reactions of the pyrolysis products to generate low molecular weight gases with calorific value of 1000–1200 kcal/nm. These gases could be used in internal combustion engines for direct power generation or in boilers for steam generation to produce power. In biomethanation, the putrescible fraction of waste is digested anaerobically (in absence of air), in specially designed digesters. Under this active bacterial activity, the digested pulp produces the combustible gas methane and inert gas carbon dioxide. The remaining digestate is a good quality soil conditioner.

Other technologies available are pelletisation, pyro-plasma, and flashpyrolysis. All these technologies have merits and demerits. The choice of technology has to be made based on the waste, quality, and local conditions. The best compromise would be to choose the technology, which (1) has lowest life cycle cost, (2) needs least land area, (3) causes practically no air and land pollution, (4) produces more power with less waste, and (5) causes maximum volume reduction.

Review Questions

1. Define sewage.
2. What is the difference between domestic sewage and industrial sewage?

45

Microbiology of Sewage

Sewage is waste water from a community containing solid and liquid excreta, derived from houses, streets and yard washings, factories and industries. It resembles dirty water with an unpleasant smell. The term *sullage* is applied to waste water which does not contain human excreta, e.g. kitchen and bathroom waste water.

The amount of sewage that flows in the sewers depend on:

- *Habit of people*   If people use more water, there will be more sewage.
- *Time of the day*   In the morning (when people tend to use more water), there will be greater quantity of sewage flowing than in the mid-day (when the flow is less) and again the flow becomes more during the evenings.

COMPOSITION OF SEWAGE

It contains 99.9% water. Solids which barely comprise 0.1% are partly organic and partly inorganic or partly in suspension and partly in solution. Offensive nature of the sewage is mainly due to the organic matter which it contains. In addition, sewage is charged with numerous living organisms derived from faeces, some of which may be agents to diseases. It is estimated that one gram of faeces may contain about 1000 million *E.coli*, 10–100 million of faecal streptococci and 1–10 million spores of *Clostridium perfringens* besides several other pathogens.

CHEMICAL CHARACTERISTICS OF SEWAGE

Domestic sewage contains 99.9% water, 0.02–0.03% suspended solids and other inorganic (30%) and organic (70%) substances. Inorganic components include ammonia, chloride salts and metals. Metal industries and mines also contribute to the inorganics. Organic components include either nitrogenous compounds like proteins and amino acids and non-nitrogenous compounds like carbohydrates and lipids. Animal sewage is high in protein and lipids and plant sewage is rich in cellulose and lignin. Lipids in the form of fatty acids which escape digestion in the digestive system account for the lipids in the faeces.

MICROBIOLOGICAL CHARACTERISTICS

The composition of microorganisms varies according to the source of waste water. Sewage water normally comprises of fungi, protozoa, algae, bacteria and viruses.

Raw sewage contains millions of bacteria/ml including coliforms, streptococci, anaerobic spore forming bacteria, *Proteus,* etc.

Soil-borne bacteria include *Bacillus subtilis*, *B. megaterium*, *B. mycoides*, *Pseudomonas fluorescence*, *Achromobacter* sp. and *Micrococcus* sp.

Bacteria of intestinal origin which are harmless include *E. coli*, *Proteus* and *Serratia* and the bacteria of pathogenic ones include enterococci, *Clostridium perfringens*, *Vibrio cholerae*, *Salmonella typhi*, *Salmonella paratyphi*, *Shigella dysenteriae*.

Viruses causing poliomyelitis, infectious hepatitis and Coxsackie, and excreted in the faeces of infected host are seen in the sewage.

TESTING OF SEWAGE

Sewage is most commonly tested for:

- total solids
- nitrogen
- chlorides
- hydrogen sulphide
- organic carbon
- dissolved oxygen
- grease and fats
- chemical oxygen demand
- biological oxygen demand

These tests are important to analyse the strength of sewage and to what extent, it can harm the environment. Presence of chemicals in sewage indicate the various treatment nature of the sewage (whether the sewage is fresh or is it decomposed anaerobically or whether it contains lot of organic matter, etc.)

DISSOLVED OXYGEN

A knowledge about dissolved oxygen in the sewage is essential from the point of view of aquatic life. Oxygen in water is available to the plants and animals that live there only if it is dissolved. Dissolved oxygen or DO can range in concentration from 0 to 14.6 parts per million in water. This is also equivalent to a weight-based measure, milligrams per litre (or mg/l). The amount of oxygen that can be dissolved in water is inversely related to temperature, i.e. as the water temperature gets higher, the amount of oxygen that can be dissolved in the water goes down. It is also possible under some circumstances to have oxygen levels above 14.6 mg/l. This can happen where water goes over a dam or other structures that causes unusual amounts of mixing. The more oxygen that is in the water, the more diversity can be expected in the plants and animals found in the water. Pollutants that make DO go down (besides heat) are organic wastes such as animal or human sewage or any chemicals that will be decomposed by bacteria in the water. The growing bacteria that break down either the organic or chemical wastes consume oxygen for their reproduction and thus deplete oxygen in the water.

Estimation of Dissolved Oxygen

The estimation of dissolved oxygen is done by titrimetric method. A known amount of the oxidising agent, manganous sulphate is added to the water to be estimated.Manganous sulphate reacts with the oxygen present in the water. During the reaction, the oxygen is bound to the manganese (chemical element Mn), forming a brownish solid which settles to the bottom of the bottle (MnO_2). This process is called fixing the oxygen. In order for this fixation process to work however, the solution must be at high pH, so another reagent is added to facilitate the above requirement. Potassium iodide is added to function as a dye, and will react with the sulphuric acid added.

Upon addition of the sulphuric acid, the MnO_2 from above is reduced to Mn^{2+}, and the iodine from the potassium iodide above is oxidised by the MnO_2 from I^- to I_2. This reaction step effectively causes the solution to take on a yellowish brown colour proportional to the number of I_2 molecules present which in turn is proportional to the original amount of O_2 molecules in the water. This iodine is titrated by sodium thiosulphate solution using starch as an indicator.

Sodium thiosulphate standard solution is added drop by drop to the mixing bottle, swirling to mix after each drop. The dropper is held vertically above the bottle and each drop is counted as it is added. As drops of this chemical enter the solution, the sodium separates from the thiosulphate ion. The thiosulphate then reacts with iodine (I_2) molecules available in the water. When the iodine molecules react, they break up into iodide- ions which are colourless.

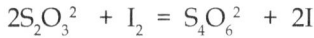

$$2S_2O_3^{2} + I_2 = S_4O_6^{2} + 2I$$

Thus 4 molecules of the sodium thiosulphate are required to change the colour resulting from one molecule of O_2 in the original water. Thus we get a very accurate estimate of the number of O_2 molecules in the original solution.

The dissolved oxygen in any water body diminishes due to various reasons which include:

■ Heterotrophic microbial activity in water.

■ Dissolved oxygen used up by aquatic microorganisms, plants and animals.

■ Addition of sewage or other organic wastes.

■ Decomposition of dead organic matter.

Review Questions

1. Write a short note on microbiology of sewage, giving highlight on its composition.

2. What are the chemical characteristics of sewage?

3. Write short notes on dissolved oxygen, throwing some light on its estimation.

46

Biochemical Oxygen Demand

Biochemical oxygen demand (BOD), sometimes referred to as biological oxygen demand, is a quantitative expression of the ability of microbes to deplete the oxygen in waste water. This depletion is caused by the microbes consuming organic matter in the water via aerobic respiration. This type of respiration uses oxygen as an electron acceptor, and the organic material being consumed provides the energy source.

If oxygen is supplied to a sewage effluent containing bacteria, aerobic decomposition will occur till the oxygen demand is satisfied. The amount of oxygen absorbed during this self-purification process is called the BOD. BOD is defined as the amount of oxygen required for the degradation of organic materials and for the oxidation of inorganic materials. It is an important indication of the amount of organic matter present in the sewage (strength of the sewage).

BOD5 TEST

The test is performed to find the amount of biologically active matter present and the rate at which the biologically active matter will be stabilised.

The BOD5 test is the primary method for determining the strength of waste water to be treated by a biological process and the strength of effluents in terms of the load on the oxygen resources of the receiving waters. The BOD test as used for assessing the efficiency of waste water treatment, is intended to be a measure of some fraction of carbonaceous oxygen demand, that is the oxygen consumed by heterotrophic microbes which utilise the organic matter of the waste in their metabolism, and not the oxygen demand exerted by autotrophic nitrifying bacteria. Since ammonia is usually present in waste waters, nitrification inhibitors must be used to suppress the exertion of nitrogenous oxygen demand. Carbonaceous oxygen demand is called first stage BOD and nitrogenous oxygen demand is called second stage BOD.

BOD is aerobic in nature. In this we can see some of the bacterial transformations related to BOD.

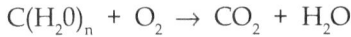

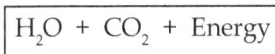

$$\text{Proteins} + O_2 \rightarrow CO_2 + H_2O$$

$$C(H_2O)_n + O_2 \rightarrow CO_2 + H_2O$$

$$\text{Fats} + \text{Oils} \rightarrow CO_2 + H_2O$$

$$\text{Organic matter} + O_2$$

| $H_2O + CO_2 + \text{Energy}$ | | New bacterial cells |

The population of bacteria first increases and later decreases because there is food problem.

PROCESSES DURING OXYGEN CONSUMPTION

The microorganisms which directly assimilate the waste organic matter may be called as primary heterotrophs. While growth is occurring, some of the organic compounds are used to synthesise compounds which are stored in the microbial cells as a food reserve. Assimilation is considered to be the first step of BOD exertion, and when most of the organic compounds have either been oxidised or converted into new microorganisms or stored food reserves, it has been completed.

The assimilation phase is followed by a period of endogenous metabolism which is manifest by oxygen consumed as the primary heterotrophs metabolise their stored food reserve. Endogenous metabolism is a mechanism by which the microbial cells meet their maintenance requirements, i.e. replace protein molecules and other cellular components which break down spontaneously.

As some of the primary heterotrophs die, the cell membranes rupture and the cell contents spill out. This organic material supports the growth of secondary heterotrophs. This growth which takes place when there is an overall decrease in the total number of viable cells present is called cryptic growth. Secondary heterotrophs also include slowly growing bacteria or fungi which are able to assimilate waste organic compounds which are not utilised by the primary heterotrophs.

In addition, protozoa and microscopic invertebrates which feed on the bacteria and fungi are also present in the waste waters. Oxygen consumption by the predators occurs while the heterotrophs are growing as well as when they are in the endogenous phase but this type of oxygen consumption is quantitatively much more significant during the endogenous phase.

Waste water contains ammonia and proteinaceous material which will release ammonia when assimilated. Ammonia serves as an energy source for nitrifying bacteria. Nitrification is an oxygen demand which is independent of the organic content of the waste water of effluents and interferes with the actual oxygen consumed by the microorganisms. Reduction of the nitrogen from the effluent is an important part of the effectiveness of waste water treatment in preventing water pollution.

If biological waste water treatment processes are exposed to sunlight they will often grow significant amounts of algae. This is more common for attached growth processes (trickling filters). Any algal cell incorporated into a BOD sample will, when incubated in the dark, consume rather than produce oxygen, and this oxygen consumption is measured as part of the BOD, but the algal presence is not related to waste organic matter originally present in the waste water and is not proportional to the ultimate carbonaceous oxygen demand of the effluent.

BOD bottles are used for testing the concentration of oxygen in a waste sample more so the amount of residual oxygen which directly relates to the presence of organic matter and microbial load in the water sample. BOD bottles are either made of glass or plastic.

The important characteristics of a BOD bottle are

- it serves as a culture vessel for the waste water microorganisms and
- it can be filled completely with a dilute sample and then sealed from contact with the atmosphere.

HOW IS BOD PERFORMED

When testing for BOD, the first step is to collect equal amounts of water from the research waterway and dilute each of the samples (usually 2 or 3 equal-sized containers) with exactly the same amount of distilled water.

The fact that many samples require dilution stems from the fact that the BOD can be higher than the available level of oxygen found in water. For example, blood discharged from a slaughterhouse can have a BOD of literally millions of milligrams per litre (e.g. BOD = 2,000,000 mg/l), which obviously would require considerable dilution before running a standard BOD. By comparison, the BOD found in raw sewage is about 200 to 250 mg/l, which would require far less dilution. Since the saturation level for the dilution water will only be 8 mg/l or so, the raw sewage would have to be diluted from 40 to 50 times to make sure it did not deplete all of the oxygen in the BOD bottle during the 5 day period i.e. one has to take care that at BOD5, residual oxygen must be present in order to calculate the difference in DO between BOD5 and BOD1 bottles.(e.g. diluting a 200 mg/l BOD solution forty-fold would drop the BOD down to 5 mg/l, which would then result in a residual DO after 5 days of about 3 mg/l).

Next, one of the samples, BOD1, is tested for dissolved oxygen content. The remaining samples are placed in foil (to keep out all light) and placed in a safe dark area where they will stay undisturbed for five days (BOD5), to seven days (BOD7), depending upon the test and time available. Then the dissolved oxygen test for BOD5 or BOD7 is subtracted from the first dissolved oxygen test for BOD1, to give the result of the Biochemical Oxygen Demand test.

Most laboratories are equipped with special BOD machines that utilise computer software to make all necessary calculations based on dilution factors, seed control, and incubation time. Otherwise standard methods provide formulae for using the DO, dilution factors, and seed control results to calculate the BOD. The decrease in the DO in a BOD bottle over a period of 5 days provides a measure of the respiration of the microorganisms present because there is no interchange with other sources of oxygen. The rate of oxygen consumption varies with temperature, and at a constant temperature, the total amount of oxygen consumed varies with time. The DO of the diluted water should be close to saturation before the dilutions are made because the respiration of some aerobic microorganisms is inhibited if the DO drops below 1.0 mg/l during the BOD test thus giving a false negative result. The residual DO in the BOD bottle after 5 days of incubation should be at least 2.0 mg/l. Since the dilution water is seeded with organisms capable of assimilating the organic matter of the waste water, there may be a small depletion of DO in 5 days even if no waste water is added. To account for this, control bottles containing only dilution water are incubated with the BOD5 test bottles and a correction is made for the DO depletion in the controls. It is important that the dilution water does not contain any constituents which are inhibitory to the microorganisms which exert the carbonaceous oxygen demand. Hence a control must be prepared from pure organic compounds usually glucose and glutamic acid.

For measurements of the BOD5 of industrial waste waters, the dilution water must be seeded with microorganisms which are capable of degrading the waste organic compounds. The seed may be obtained from a biological treatment process which has been applied to the waste water or from sediment downstream from the point at which the effluent is discharged into the receiving waters. If a diluted sample of waste water is incubated under the proper conditions and the oxygen consumption is recorded every day, the rate of oxygen consumption is high initially but decreases continuously over time, approaching zero finally.

Microbiological Phenomena
Which Result in Oxygen Consumption

An assumption that the rate of microbial oxygen consumption is proportional to the concentrations of ultimate carbonaceous BOD cannot be made since there are different microbiological phenomena which occur in a BOD bottle

- assimilation of organic matter
- endogenous respiration
- cryptic growth
- oxygen consumption by predators.
- oxygen consumption by nitrifying bacteria
- oxygen consumption by algae

The BOD bottles and a BOD incubator are shown in Fig. 46.1.

CALCULATION OF BOD

$$BOD5 = [(C1 - C2) - ((Vt - Ve)/Vt)(C3 - C4)] (Vt / Ve)$$

where

- C1 is the dissolved oxygen concentration, in milligrams per litre, of the test solution at time zero.
- C2 is the dissolved oxygen concentration, in milligrams per litre, of this same solution after five days.
- C3 is the dissolved oxygen concentration, in milligrams per litre, of the blank solution (diluent) at time zero.
- C4 is the dissolved oxygen concentration, in milligrams per litre, of the blank solution (diluent) after five days.

- V_e is the volume, in millilitres, of sample used for the preparation of the test solution concerned.

- V_t is the total volume, in millilitres, of this test solution.

Things to Remember during BOD Determination

- All samples must be kept at or below 4°C until the sample is ready to be tested.

- Because microbes are most active between pH values of 6.5 and 7.5 all samples must be neutralised to a pH within this range.

- It is also important to remove any residual chlorine from the sample before running the experiment. Failure to do so would kill the valuable bacterial source.

BOD bottles   BOD incubator

Figure 46.1   *Equipment for BOD Determination*

- A seed source (bacterial control) must be prepared from a commercial product, or from a diluted portion of waste water. This seed source will be the source of bacteria added to the sample. The bacteria are added to oxidise the biodegradable organic matter in the sample. This is the backbone of the experiment. If bacteria fail to oxidise the organic matter, then the test will not work. It is also important to run a seed control to account for any BOD in the seed material.

- Because the BOD concentration in most waste water exceeds the concentration of DO available in air saturated samples, it is necessary to dilute the samples being tested. This is done by using properly prepared dilution water. This dilution water contains nutrients such as nitrogen, phosphorus, and other trace metals necessary for microbial growth. The dilution water is also buffered to ensure that the pH of the sample remains in a suitable range for microbial activity.

Review Questions

1. Define BOD.
2. Describe the process of BOD estimation.

47

Chemical Oxygen Demand

Chemical oxygen demand is another means of measuring the strength (in terms of pollution) of waste water. By using this method, most oxidisable organic compounds present in the waste water sample may be measured. COD measurements are preferred when a mixed domestic-industrial waste is entering a plant or where a more rapid determination of the load is desired.

COD is the amount of oxygen required for the chemical oxidation of organic matter with the help of strong chemical oxidants.

The COD test measures not only the oxygen equivalent of the waste organic matter but also that of the microbial cells. The oxygen demand associated with the microbial cells is only partially exerted during a BOD test, also some of the organic compounds measured by the COD determination may not be metabolised by the microorganisms in either the BOD bottle or the biological treatment process. There are many different analytical approaches to quantify the presence of oxidisable contaminants in waters. Over a century ago, a potassium permanganate solution (i.e. a strong, dark purple coloured oxidant) was added to waters. In turn, if there were readily oxidisable materials present, the colour of this solution would progressively turn less purple, and perhaps even turn completely clear if all of the oxidising permanganate were to be completely reduced.

The permanganate test was more so qualitative than quantitative, and it was eventually replaced with a comparable chemical analysis which we call the chemical oxygen demand (COD). In this case, though, the oxidising agent is potassium dichromate (i.e. a strong, dark orange coloured oxidant). The dichromate is mixed with the sample, and then by the additional use of a strong acid (i.e. sulphuric acid) and heat, the chemical reduction of the dichromate takes place parallel to the oxidation of the existing contaminants. This reaction is allowed to continue for 90 minutes, at which time the sample is cooled and analysed for the residual presence of the dichromate using a colorimetric method (i.e. where the orange colour gives way to a kelly-green colour as the oxidised chrome (hexavalent chrome) reduces to the reduced form (trivalent).

The organic matter is of two types:

- that which is oxidised biologically and called biologically active.
- that which cannot be oxidised biologically and called biologically inactive.

By performing a COD one gets an idea about the total or entire organic matter and actually what one is interested to find out is the biologically active matter which becomes the basis for further treatment.

COD has certain advantages:

- COD values for a given sample will be greater than BOD. The reason is that biochemical oxygen demand measures only the quantity of organic material capable of being oxidised, while the chemical oxygen demand represents a more complete oxidation.
- Results of COD can be obtained within 5 hours as compared to 5 days of BOD.
- COD procedure is relatively easy and can give reproducible results.

The following ranges for COD results are given for general reference and apply primarily to average domestic waste water. Significant amounts of industrial waste discharges may cause wide variations in these ranges.

Plant Influent   →   300–700 mg/l

Primary Effluent   →   200–400 mg/l

Trickling Filter Effluent   →   45–130 mg/l

Activated Sludge Effluent   →   30–70 mg/l

Advanced Waste Treatment Effluent   →   5–15 mg

Review Questions

1. What is COD?
2. Explain the process of COD.

48

Sewage Treatment

Wastes are treated by a variety of sewage treatment processes that are aimed at reducing the biological oxygen demand and removing nutrients that could cause eutrophication of receiving waters. Sewage treatment can involve physical removal of solids (primary treatment); biological decomposition of organic compounds (secondary treatment); chemical, physical or biological removal of other constituents such as heavy metals, nitrogen, and phosphates (tertiary treatment); and disinfection to remove potentially pathogenic microorganisms.

NEED FOR SEWER SYSTEM

Each time you flush the toilet or you wash something down the sink's drain, you create sewage (known as waste water). There are three main reasons why waste water cannot be directly released into the environment. First of all, it creates bad odour, secondly it contains harmful bacteria which cause health hazards and thirdly, it contains suspended solids and chemicals that affect the environment. That is why communities build waste water treatment plants and enforce laws.

The aim of sewage purification is:

- To stabilise the organic matter i.e. to break down the organic matter to simple substances which will not decompose further. This is accomplished by bacterial action.

- To produce an effluent which is free from pathogens and which can be disposed off without causing a nuisance.

- To utilise the water and solids, if necessary, without risk to health.

DECOMPOSITION OF ORGANIC MATTER

The decomposition of organic matter in sewage (stabilisation of sewage) takes place by two processes: aerobic and anaerobic.

Aerobic process   It is the most efficient method of reducing the organic matter in sewage. The process requires a continuous supply of free dissolved oxygen. The organic matter is broken down into simpler compounds by the action of bacteria, fungi and protozoa.

Anaerobic process   When the sewage is highly concentrated and contains plenty of solids, the anaerobic process is highly effective. In anaerobic decomposition, the reactions are slower and the mechanism of decomposition is extremely complex.

Sewage treatment is managed on small scale in individual homes and rural areas as well as on large scale in towns by municipal bodies.

SMALL SCALE SEWAGE TREATMENT

This essentially involves the sewage treatment in unsewered areas.

Cesspools

In many houses, human waste is dumped into cesspools. Cesspool is an underground construction consisting of concrete cylindrical rings with pores in the walls of the ring. Water passes out the bottom and through the pores into the surrounding sand whereas the solid waste accumulates at the bottom. Anaerobic bacteria in the sludge layer at the bottom of cesspool digest the organic matter. The breakdown products diffuse into the ground. After few years, if the sludge becomes thicker it is necessary to clean with strong acid. Dried bacterial spores of *Bacillus subtilis* are available in stores. These may be added to accelerate the sludge digestion. Addition of yeasts at intervals is also useful.

Septic Tank

It is a water-tight masonry tank into which household sewage is admitted for treatment. It is a satisfactory means of disposing excreta and liquid wastes from individual dwellings, small groups of houses and institutions which have adequate water supplies but do not have access to public sewerage system. Fig. 48.1(a) shows the location of a septic tank in a household. Septic tanks are settling tanks and are based on the sedimentation principle. They are plain single storeyed tanks Fig. 48.1(b).

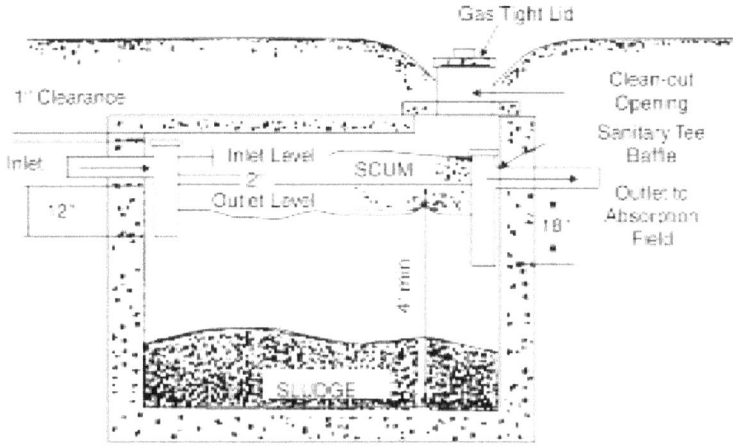

Figure 48.1(a) *Structure of a septic tank*

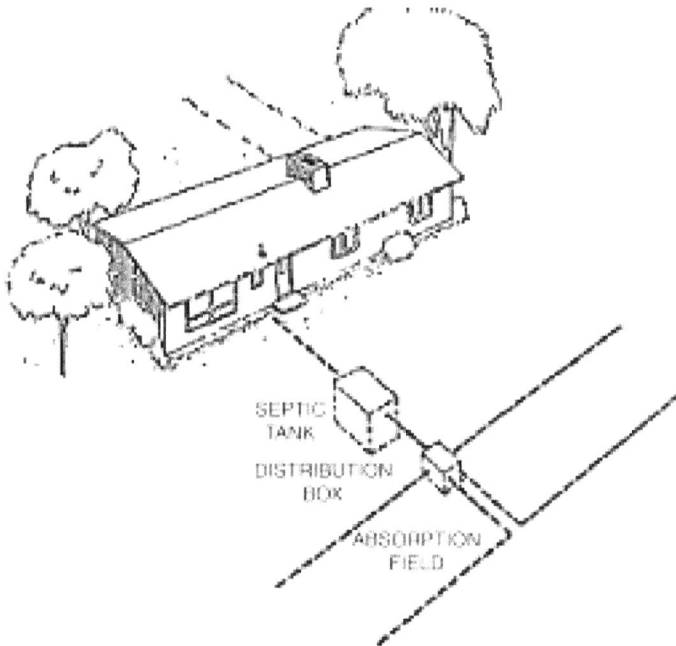

Figure 48.1(b) *Location of septic tank in a household*

It is the simplest form of anaerobic treatment system. It is a kind of sewage settling tank. It is an enclosed concrete box into which waste flows from the house. The organic matter accumulates at the bottom of tank whereas the water rises to the outlet pipes which empty into the surrounding area. The tank is to be pumped out regularly as there is no absorption of the digested organic matter into the earth. Complex organic matter are anaerobically decomposed to simple organic molecules and fermentation gases. They are relatively small closed tanks vented for the escape of fermentation gases. The tank accomplishes two processes¾sedimentation and biological degradation of the sludge which are achieved by anaerobic digestion and aerobic oxidation.

Anaerobic digestion and aerobic oxidation   The human excreta consists of 65% mineral matter (which do not undergo any chemical changes in a septic tank) and 35% organic matter of which only 20–40% of organic matter (solids) are liquified or gasified in the septic tank. The heavier matter (sludge) settles at the bottom and the lighter matter (grease and fats) forms a layer called scum on the top. Solids are attacked by the bacteria and are broken down into simpler compounds. This is the first stage of purification called anaerobic digestion. The sludge is much reduced in volume and is rendered stable and inoffensive. A portion of the solids is transferred into liquid and gases (mostly methane) which rises to the surface in the form of bubbles. The liquid which passes out of the outlet pipe from time to time is called effluent. It contains numerous bacteria, cysts, helminthic ova and organic matter in solution or fine suspension. The effluent is allowed to percolate into the subsoil. There are millions of aerobic bacteria in the upper layers of the soil which attack the organic matter present in the effluent. Thus the organic matter is oxidised into stable end products like nitrates, carbon dioxide and water. This stage of purification is the aerobic oxidation. Thus two stages are involved in the purification of sewage: First stage (anaerobic digestion) takes place in the septic tank and the second stage (aerobic oxidation) takes place outside the septic tank in the subsoil. Sedimented sludge accumulated at the bottom of septic tank is removed and may be dried, ground and spread on soil as a fertiliser. In a typical septic tank, about 80% of the incoming suspended solids will be removed and the BOD will be reduced. However, this method cannot be relied upon to eliminate pathogenic microorganisms in sewage. The effluent will contain large number of bacteria, and discharge from a septic tank may contain even *E.coli*/100ml. Hence it is important that the tank should not be in the proximity of water supplies.

Imhoff's Tank

This is an improved design of a septic tank that maintains anaerobic condition more strictly, produces some utilisable biogas and facilitates the settling of solids, but requires expert maintenance. The tank is divided into two sections (Fig.48.2), thus they are double storeyed settling tanks. The upper section acts as the settling tank. Fresh sewage is introduced in the upper compartment which slips by an opening in the lower compartment. The settling compartment is separated from the digestion compartment. Solids move down into the digestion compartment but are prevented from being carried back up into the settling compartment. Stabilised solids (digested) are removed from the bottom of the tank periodically. Usually, Imhoff tanks are 9 m deep and their construction is costly. The scrap, grease, oil, etc. flood at the top and form a scum. Imhoff tank is quite efficient in allowing complete digestion in 2–4 hours. A diagrammatic representation of the Imhoff's tank is shown in Fig. 48.2.

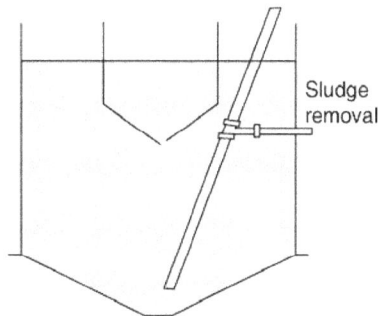

Figure 48.2 *A diagrammatic representation of the Imhoff's tank*

LARGE SCALE SEWAGE TREATMENT (SEWAGE TREATMENT IN SEWERED AREAS)

Urban areas (municipal corporations) which are sewered areas have to handle large proportions of wastes and are quite complex. Large-scale sewage treatment involves three processes namely (a) primary/preliminary/physical treatment (b) secondary/biological treatment (c) tertiary /chemical treatment

PRIMARY/PHYSICAL/PRELIMINARY TREATMENT

The primary treatment process consists of screening and sedimentation processes.

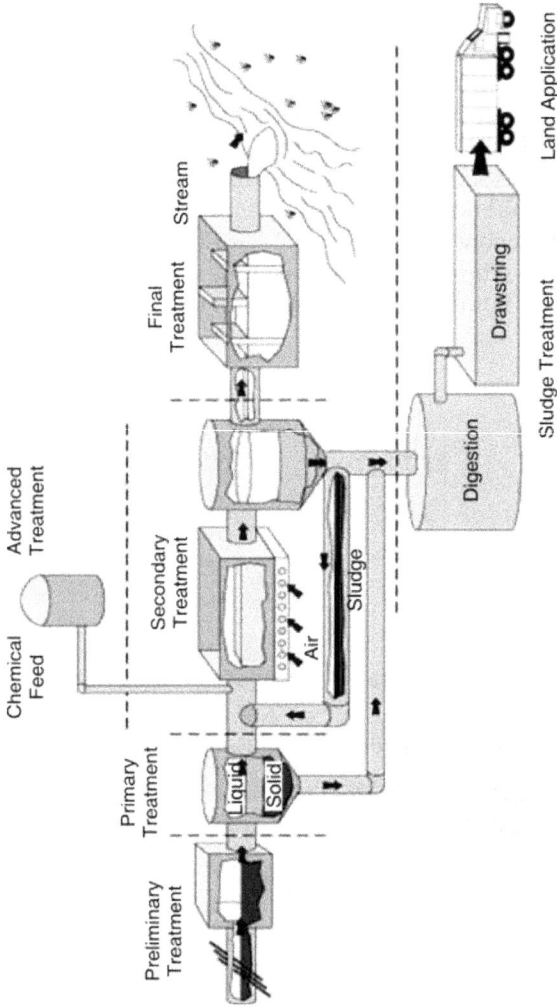

Figure 48.3   *Schematic representation of large-scale sewage treatment*

Screening

The purpose of screening is to remove larger floating solids and larger organic solids which do not aerate and decompose (become septic), and to skim grease and fatty oils.

This process is accomplished by screeners which are of three types:

- Rock or coarse screens which are made of parallel bars 75 mm apart. It involves more area and hence is more effective.
- Medium screens having bars or mesh with a width of 12.5 - 40mm.
- Fine screens made of fine wire of perforated metal with openings less than 13 mm. They get easily clogged.

To prevent overloading with collected solids, the screen is slowly rotated so that solids can be removed at regular intervals. Screenings are usually disposed by burial or incineration or they are passed through a macerator which shreds them to small sizes.

Disposal of screenings

- By disintegration after which they are either dumped into sea or buried
- By returning the screening back into the sewage after maceration.
- By burying the screenings in shallow trenches covered with dry earth. This method requires lot of land.
- By comminuters (mechanical appliances used in USA) which combines a screen and a disintegrator in a single mechanism. Both are done in closed chamber.
- By mixing the screenings with house refuse composting.
- By incineration (this method is costly).

Removal of grit (Grit/Detritus Chamber)   The detritus chamber removes heavy inorganic matter like grit, sand, gravel, road scrapings and ashes. These particles are discrete particles that do not decay but create nuisance. They may injure pumps and make sludge digestion difficult. Grit particles are of large size and hence high density compared to organic matter. Thus, they are removed by differential settling. Grit particles (0.20 mm) settle at a velocity of 1.2 m/min whereas suspended solids (faeces) have considerably lower settling velocities. The grit settles 30 cm in 16 sec, if the velocity is 30 cm/sec and the detention period is 1 min. Length of the tank should be 18 m.

Storage space is provided at the bottom of the tank such that the clearing period in average is 2 weeks.

Removal of fatty oils (skimming tanks) Sewage contains lot of grease and fatty oils (which forms a scum in sedimentation tanks and interferes with oxidation process in aeration tanks) which are removed. Skimming tanks or primary settling basins (Fig. 48.4) are about 1 m deep and the scum accumulations are removed manually or buried or burnt. This method is not of much use in India.

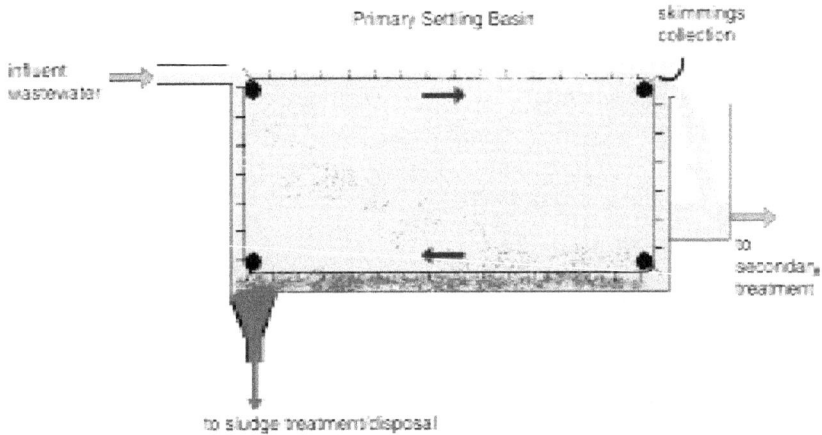

Figure 48.4 *Primary settling basin*

Sedimentation (primary sedimentation) Sewage from the primary settling basin is now admitted into a huge tank called the primary sedimentation tank (Fig. 48.5).

It is a very large rectangular tank. Nearly 50–70% of the solids settle down under influence of gravity. A reduction between 30–40% in the number of coliforms is obtained. The organic matter which settles down is called sludge and is removed by mechanically operated devices (scrappers) (Fig.48.6) without disturbing the operation in the tank. Simultaneously a small amount of biological action also takes place in which microbes present in the sludge attack complex organic solids and break them down to simpler soluble substances and ammonia. A small amount of fat and grease rises to the surface to form a scum which is removed from time to time and disposed. Purpose of sedimentation is to remove suspended solids and thereby reduce the strength of the sewage. Sedimentation or clarification is the settling and removal of suspended impurities which occur when water stands still or flows slowly through a basin or tank. There is negligible turbulence and hence particles of high density tend to settle down (gravity) and form a sludge layer at the bottom of the tank whereas clarified water will be collected through the outlet. Hence sedimentation units play dual role:

Figure 48.5 *Primary sedimentation tank*

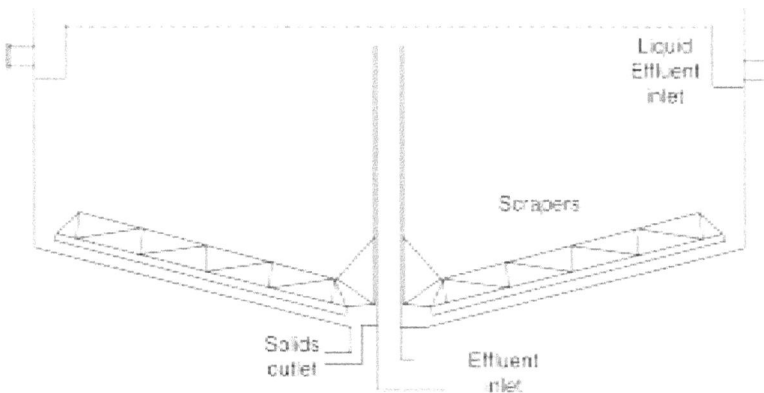

Figure 48.6 *Diagrammatic illustration of primary sedimentation tank*

- Removal of settleable solids.
- Concentration of removed solids into sludge.

There are three types of sedimentation tanks:

- Vertical flow tanks
- Horizontal flow tanks (most common)
- Circular flow tanks or radial flow tanks which receive sewage in the centre and their flow is towards the sides.

SECONDARY TREATMENT

The effluent from the primary sedimentation tank, still contains a proportion of organic matter in solution or colloidal state and numerous living organisms. Its BOD is high hence it is subjected to aerobic oxidation by various methods. It is necessary to oxidize the organic matter present in the effluent before it is being discharged into the body of water (diluting). This oxidation is either carried out on land (naturally) or in bacterial beds (artificially) such as trickling filter or aerators. The preliminary treatment reduces the BOD while the secondary treatment satisfies the demand of BOD with the help of bacterial beds. Secondary treatment relies on microbial activity. Since these oxidation processes are carried out by microorganisms, they are called as 'biological oxidators' There are four different types of oxidation processes:

Fixed film sewage treatment

- Biological filters or trickling filters
- Rotating biological contractor or biodisc system

Suspended cell sewage treatment

- Oxidation ponds
- Activated sludge process

Trickling Filter System

This is a fixed film sewage treatment method (Fig. 48.7). It is a simple method but very expensive. It works on the principle of filtration. Sewage is distributed by a sprinkler revolving over a bed of porous material which is normally an artificially constructed bed consisting of broken stones, bricks or other suitable material. The aerobic life that flourishes on the surface of such material oxidises and nitrifies the organic matter present. Whenever sewage flows over a contact surface, aerobic bacteria lying dormant in the liquid, become active and start breeding readily in favourable conditions and form a film called the zooglial layer on the surface. Sewage percolates the porous bed and the effluent is collected at the bottom. Dense slimy bacteria grow and coat the porous material. *Zooglea ramigera* has a principal role in generating slime matrix through secretion of exopolysaccharide which accumulates a heterogeneous microbial community (bacteria, fungi, protozoa, nematodes and rotifers). This community absorbs and mineralises the dissolved organic nutrients in the sewage thus reducing the BOD of the effluent. Aeration is passively provided by the porous material. A food web is established based upon the microbial biofilm. Insects consume the excess biomass generated (only when the sprinkling is shut off). Sewage is re-circulated several times through the same filter to further clear the sewage.

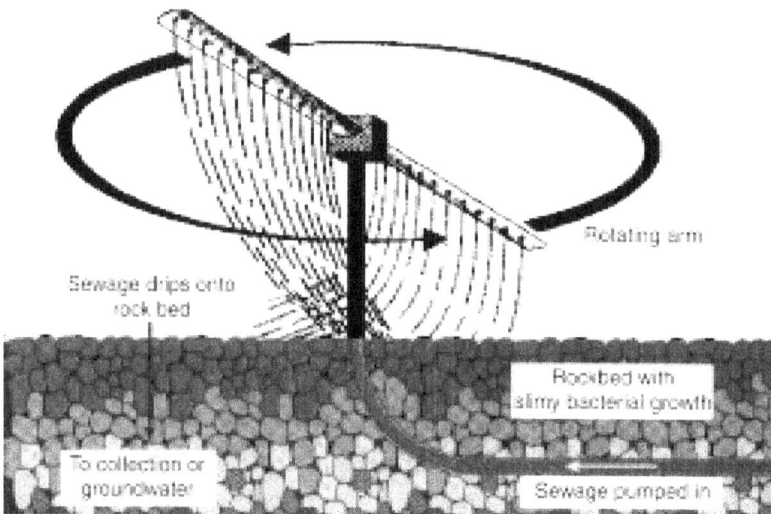

Figure 48.7 *Trickling filter*

Drawbacks Nutrient overload may lead to excess microbial slime reducing aeration and percolation rates and leads to removal of trickling filter bed. Trickling filter method cannot be used during cold winters when the temperature is very low since the growth rate of the organisms becomes very low.

Rotating Biological Contractor or Biodisc System

This is also a fixed flow sewage treatment method. Closely spaced discs manufactured from plastic material are rotated in a trough containing sewage effluent. The discs are partially submerged in effluent and they become coated with a microbial slime. Continuous rotation keeps the slime well-aerated and in contact with the sewage. Biofilms develop on the rotating biological contractors, and among the filmamentous microorganisms that are found are *Sphaerotilus*, *Beggiatoa*, *Nocardia* and *Oscillatoria*. The rotation bio-contractors biofilm has two layers: outer whitish *Beggiatoa* filament and inner blackish *Desulfovibrio* (anaerobic) layer.

Figure 48.8 *Rotating Biological Contractor*

The inner layer is blackish due to the precipitation of iron sulphide when iron reacts with hydrogen sulphide produced by sulphate reducing *Desulfovibrio* species. In the anaerobic zone, fermentative microbes produce organic acids and alcohols that are used up by *Desulfovibrio*. Hydrogen sulphide produced by the *Desulfovibrio* diffuses into the outer aerobic zone where it is utilised by *Beggiatoa* as an electron donor. Hydrogen sulphide is oxidised by *Beggiatoa* and the elemental sulphur formed, gets deposited within the cells of bacteria (Beggiatoa). Thickness of microbial slime layer in all film flow processes is governed by the diffusion of nutrients through the film. When the film grows to a thickness that prevents nutrients from reaching the innermost microbes, these will die or autolyse and cause detachment of slime layer which can be easily removed by settling.

Advantages

- Requires less space than trickling filters.
- More efficient and stable.
- Produce no aerosols.
- Successful in treating both domestic and industrial sewage.

Disadvantage

- Needs higher initial investment.

Oxidation Lagoons

Oxidation ponds (Fig. 48.9) are also known as stabilisation ponds or lagoons. They are used for simple secondary treatment of sewage effluents. Within an oxidation pond, heterotrophic bacteria degrade organic matter in the sewage which results in production of cellular material and minerals which supports the growth of algae in the oxidation pond. Growth of algal populations allows further decomposition of the organic matter by producing oxygen. The production of this oxygen replenishes the oxygen used by the heterotrophic bacteria. Typically oxidation ponds need to be less than 10 feet deep in order to support the algal growth. In addition, the use of oxidation ponds is largely restricted to warmer climatic regions because they are strongly influenced by seasonal temperature changes. Oxidation ponds also tend to fill (become shallow), due to the settling of the bacterial and algal cells formed during the decomposition of the sewage. Overall, oxidation ponds tend to be inefficient and require large holding capacities and long retention times. The degradation is relatively slow and the effluents containing the oxidised products need to be periodically removed from the ponds.

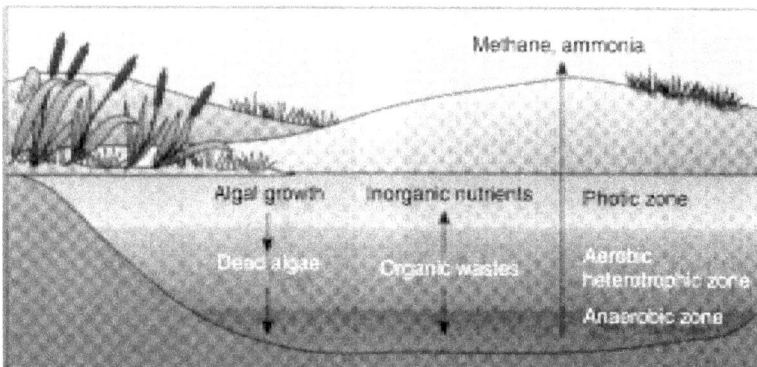

Figure 48.9 *Oxidation ponds*

Microbes grow as suspended particles within the water column rather than as biofilms. As oxygenation is usually achieved by diffusion, and by the photosynthetic activity of algae these systems need to be shallow. Sewage is subjected to primary settling and is subsequently channelled through a series of oxidation ponds. Treated sewage is discharged into surface waters. Partially treated sewage is used for ground water recharge. Primary pond re-aerates the oxygen-depleted water. After settling of most of the algal and bacterial biomass, the water is transferred to large shallow infiltration ponds. From these ponds, water flows through sand and soil layer and slowly returns underground. Clogging of infiltration ponds (undegraded microbial polysaccharide and accumulation of iron sulphide) is a recurrent problem in ground water treatment operation. Problem is countered by periodic rest periods for infiltration basins which allow degradation of the excess polysaccharides and re-aerates the sediment of oxidation pond with the oxidation of clogging iron sulphide.

Activated Sludge Process

This is the most common option in secondary treatment and is also a method of suspended cell sewage treatment method. It starts with aeration that

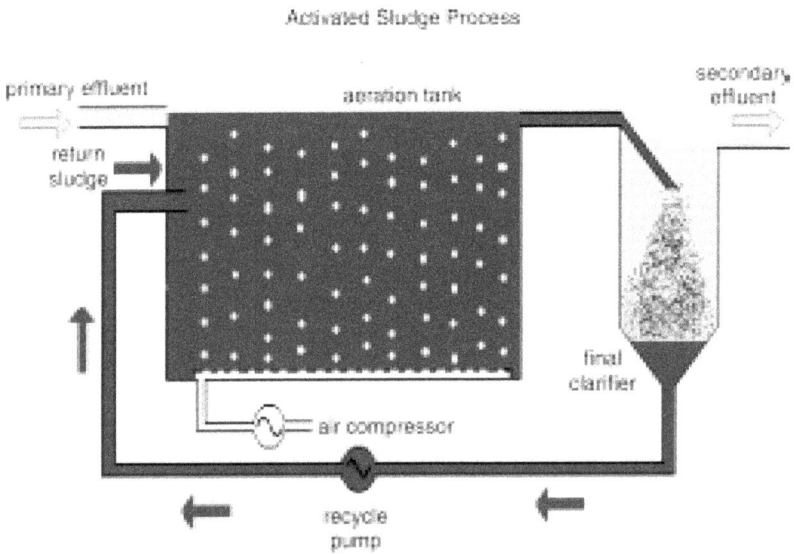

Figure 48.10 *Activated sludge process*

encourages the growth of microbes in the waste. The microbes feed on the organic material, which then allows solids to settle out. Bacteria-containing 'activated sludge' is continually re-circulated back to the aeration basin to increase the rate of organic decomposition. After primary settling, sewage containing dissolved organic compound is introduced into an aeration tank. Aeration is provided by air injection or mechanical stirring by allowing compressed air continuously from the bottom of the aeration tank.

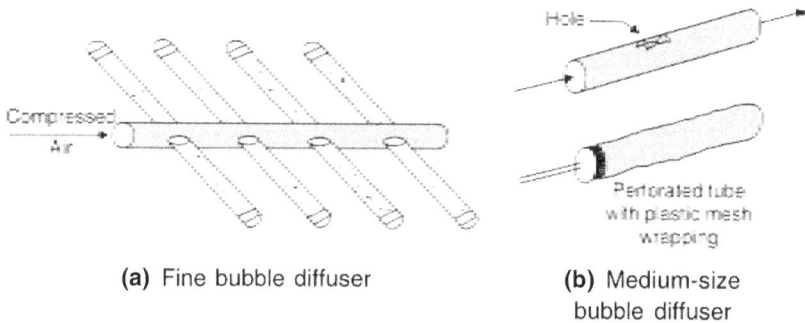

(a) Fine bubble diffuser

(b) Medium-size bubble diffuser

Figure 48.11 *Release of air from the bottom of tank*

Microbial activity is maintained at high levels by introduction of most of the settled activated sludge (also called returned sludge which is rich in culture of aerobic bacteria) thus the name activated sludge process. During the process of aeration, organic matter of the sewage gets oxidised with the help of aerobic bacteria in the tank which are suspended in the sewage water. The typhoid and cholera organisms are definitely destroyed and the coliforms are greatly reduced. During the holding period in the tank, there is vigorous development of diverse heterotrophic bacterial population like *Micrococcus, Achromobacter, Athrobacter, Flavobacterium, Zoogloea*. Filamentous bacteria like *Sphaerotilus,* mycobacteria are also common. Filamentaous fungi and yeast occur in low numbers. Protozoa are represented by ciliates. They are important predators of bacteria.

Bacteria occur individually in free suspension aggregates as floccules by *Zoogloea ramigera*. Floc is too large to be ingested by protozoa hence considered as a defence mechanism. In raw sewage, bacteria predominate during the holding time in the aeration tank. The number of suspended bacteria decrease but those associated with floc greatly increase in number. During the holding period, a portion of dissolved organic substrate is mineralised. Another portion is converted to microbial biomass. In the advanced stage of aeration, most of the microbial biomass is associated with floc that can be removed from suspension by settling.

Poor settling characteristics are associated with the bulking of the sewage sludge, a problem caused by proliferation of filamentous bacteria like *Sphaerotilus, Beggiatoa, Thiothrix* and *Bacillus*, filamentous fungi such as *Geotrichum, Cephalosporium, Cladosporium* and *Penicillium*. Bulking may be caused by a high C : N and C : P ratios and low dissolved oxygen concentration. A portion of the settled sewage is recycled for inoculation of incoming raw sewage. Excess sludge requires incineration or addition of treatment by anaerobic digestion and composting or disposal as landfills. ASP tends to reduce the BOD to 5–15% of raw sewage. Treatment drastically reduces the number of intestinal pathogens in sewage (through competition, adsorption, predation and settling). Numbers of *E.coli* and enteroviruses are 90–99% lowered in the effluent by the ASP than in the incoming raw sewage. ASP is essentially a continuous culture process. In a steady stable ASP, growth of sludge bacteria (floc) must be equal to the sludge wastage. The presence of attached ciliates is critical for the control of disposed bacteria and removal of portion of BOD. The types and number of protozoa associated with the floc can be used as an indicator of sludge condition and thus treatment performance can be monitored. Sludge is in poor condition when few ciliates and many flagellates are seen. Ciliates predominate in good sludge. ASP is efficient and flexible and is able to withstand variation in sewage flow rate and concentration and is widely used for the treatment of domestic waste and industrial effluent. It produces large volume of sludge.

Modification of ASP

This process was developed by Union Carbide and uses oxygen instead of air for aeration. Closed tanks and specially designed stirring system prevent the wastage of oxygen. Deep shaft process developed by Imperial Chemical Industry uses injection of oxygen and achieves high oxygen solubility by increasing hydrostatic pressure through a vigorous circulation pattern. Some more advancements in the ASP have been developed that include anaerobic stages also. In the anaerobic stage, dilute sewage passes through a dense stationary anaerobic biomass which minimises reactor volume and maximises the efficiency of digestion process. Bacteria involved tend to form granules. Besides reducing the BOD, advanced activated sludge secondary treatment (ASST) also removes inorganic nitrogen and phosphorous which when present in the sewage effluent is concerned with the phenomenon of 'eutrophication' of natural waters. Eutrophication is a process where water becomes rich in nutrients and is a natural sequence in aquatic environment. Deep, cool, nutrient-poor water becomes shallower, warmer and more eutrophic and settles, and organic matter accumulates from biological production. Eventually, lake is filled in. Sudden nutrient enrichment through sewage discharge triggers explosive algal blooms. As a result there is mutual shedding of microorganisms and exhaustion of micronutrients. Toxic products increase eventually and algal population crashes. Decomposition of dead algal biomass by heterotrophs exhausts the dissolved oxygen in the water, leading to septic conditions.

Many natural oligotrophic lakes are limited by P/N but if the sewage effluent supplies large amount of these inorganic nutrients, eutrophication response is likely to occur. Obvious remedy is to remove the causative inorganic nutrients from the sewage discharges.

Biological Processes for Removal of Nitrogen by ASP

It involves a series of aerobic and anaerobic treatment stages. Prolonged aeration is necessary to convert the nitrogen-containing wastes to nitrates. After conversion to nitrate, it is possible to remove most of the nitrogen from the effluent by creating anaerobic condition for denitrification. Hence aeration is discontinued and organic matter is added. Anaerobic oxidation of organic matter converts the nitrate to nitrogen gas which escapes into the atmosphere.

Removal of Excess Phosphates

Bacterial uptake and storage of P in the form of polyphosphates may be encouraged by modifications in the treatment processes. Enhanced phosphate removal from sludge is associated with polyphosphate storage by *Acinetobacter*, *Moraxella mima* group of bacteria. Under anaerobic condition, the microbes incorporate large amounts of fatty acid and store them in the form of Poly Beta Hydroxybutyrate (PHB). When the sewage is re-aerated, the polyphosphate bacteria rapidly oxidise their intracellular PHB and take in the phosphate that is in excess of their growth requirements. The incorporated phosphate is stored as energy rich polyphosphate in cells thus removing phosphorus from the liquid effluent. All the advanced ASP involves one or more alternate oxic or anoxic sewage storage cycles. Anaerobic or oxidation process includes anaerobic zone with a retention time of 0.5–1 hour and an aerobic tank of retention time of 1–3 hours. During anaerobic phase, inorganic phosphorous is released from the cells as a result of polyphosphate. Energy liberated is used for the uptake of BOD from waste water. During the aerobic process, soluble phosphate is taken up by the bacteria which synthesise polyphosphate using the energy released from BOD oxidation. The A/O process results in the removal of phosphorous and BOD from effluent and produces a rich sludge.

Submerged Aerobic Filters

This system consists of partially or totally submerged biological support material and improves the efficiency of biological filtration. In submerged or flooded aerobic filter, effluent flows down through a bed of small natural or plastic media. Air is injected into the base of bed. Hence waste water can be applied at a much higher rate than conventional filters. There is a second filtration stage below the injection point where removal of solids generated in the upper aerobic mixed zone occurs. Solids, on accumulation in the lower filter stage, reduce a critical stage in clearing by backwashing. Wash water is returned to the treatment plant for settlement.

Sedimentation (Secondary Sedimentation)

The oxidised sewage from the trickling filters or aeration chamber is led into the secondary sedimentation tank (SST) where it is detained for 2–3 hours. The sludge that collects in the SST is called aerated sludge or activated sludge because it is fully aerated. It differs from the sludge in the primary sedimentation tank in that it is practically inoffensive and is rich in bacteria, nitrogen and phosphates. It is a valuable manure, if dehydrated. Part of activated sludge is pumped back into the aeration tanks in the activated sludge process and the rest is pumped into the sludge digestion tanks for treatment and disposal.

Sludge Digestion

One of the greatest problems associated with sewage treatment is the treatment and disposal of the resulting sludge. One million gallons of sewage produces 15–20 tons of sludge. Sludge is a thick black mass containing 95% water and it has a revolting odour.

Some methods of sludge disposal are digestion, sea disposal and land disposal (composting).

Anaerobic digestors (sludge digestion) These are used to process the sludge. Generally slower but saves energy compared to treatment processes requiring forced aeration. Some anaerobic treatment systems also generate biogas which is a useful fuel. The simplest anaerobic system is the septic tank. Large scale anaerobic digestion is used for further processing the sewage sludge produced by primary and secondary treatments. Hence they are used for processing settled sewage sludge and for treatment of some very high BOD industrial effluents. Conventional anaerobic digestors are large fermentation tanks (Fig. 48.12) designed for continuous operation under anaerobic condition. The digestor contains large amounts of suspended organic matter (20–100g/l) with a considerable part of this suspended material forming the bacterial biomass. Fungi and protozoa do not play a significant role in anaerobic digestion.

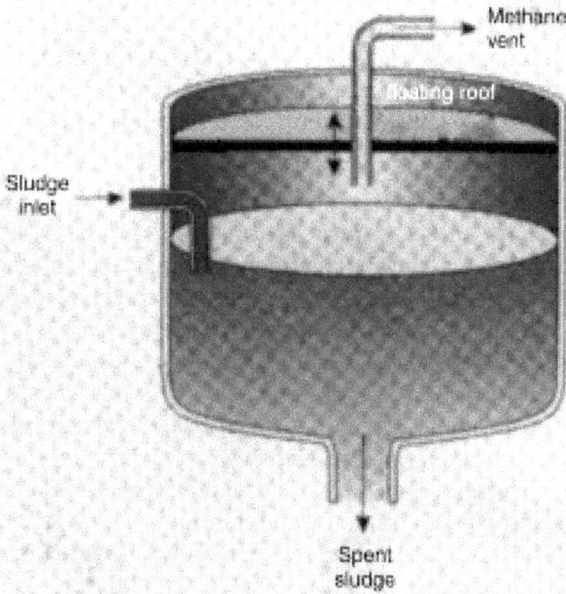

Figure 48.12 *Anaerobic sludge digestor*

Anaerobic digestion involves two steps:

1. Complex organic materials (microbial biomass) are de-polymerised and converted fermentatively to fatty acids, carbon dioxide and hydrogen (non-methanogenic anaerobic bacteria).

2. Methane is generated by the direct reduction of methyl groups to methane or by reduction of CO_2 to methane by molecular hydrogen or by other reduced fermentation products such as fatty acids, methanol or carbon monoxide. Some examples are *Methanobacterium bryantii*,

M. thermoautotrophicum, M. mobile and *M. formicicum.* It takes 3–4 weeks for complete sludge digestion and the residue is inoffensive, sticky and tarry mud which will dry readily and form an excellent manure.

Factors important for anaerobic digestion A temperature of 35–37°C and pH 6–8 is optimal. Extremes of pH, influx of heavy metals, solvents, other toxic materials can easily upset the operation of the anaerobic digestion when it becomes 'stick/som' where methane production is interrupted and fatty acid and other fermentation products accumulate. Reactor must be cleaned and charged with large volumes of anaerobic sludge or appropriate methanogenic bacteria from an operational unit. A normally operating anaerobic digestion yields a reduced volume of sludge which yields odour problems. Aerobic composting of the sludge consolidates the sludge and renders it suitable for disposal or for use as soil conditioner. Biogas produced consists of CO_2, CH_4, N_2 and H_2S and occasionally traces of H_2. Anaerobic digestors are expensive to construct and require expert maintenance.

TERTIARY TREATMENT

All the preliminary and secondary treatment reduce the BOD levels of the sewage. The aim of the tertiary treatment is to remove non-biodegradable toxic organic pollutants such as chlorophenols, polychlorinated biphenyls and other synthetic pollutants. They are removed by activated carbon filters. Phosphate is removed by precipitation as calcium phosphate. Nitrogen is removed by volatilisation as ammonia. Ammoniacal nitrogen can also be removed by breakpoint chlorination by adding hypochlorous acid in 1:1 ratio. Removal of ammoniacal nitrogen lowers the BOD because nitrification would consume oxygen dissolved in the remaining water. Removal of heavy metals like mercury, lead, chromium and cadmium also occurs during tertiary treatment. The absorbed metal ions are generally converted into either toxic products or residues that remain associated with the microbe biopolymer matrix and are either released during sludge treatment or are remobilised after sludge disposal. The general tendency of bacteria to concentrate heavy metals in their biomass is favourable to effluent quality, but it complicates the disposal of sludge (rectified to some extent by microbial mining. e.g. acid produced by thiobacilli would solubilise heavy metals and leach them from sludge). Heavy metals can be subsequently fed and reprocessed for use or permanently immobilised.

Disinfection is the final step in tertiary treatment. This is to kill escaped bacteria or viruses. This is accomplished by chlorination (chlorine gas, hypochloride a sodium hypochlorite).

$$Cl_2 + H_2O \rightarrow HOCl + HCl$$
Hypochlorous acid

$$CaO(Cl)_2 + 2H_2O \rightarrow 2HOCl + Ca(OH)_2$$
Calcium hypochlorite

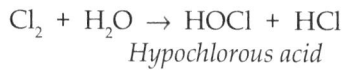

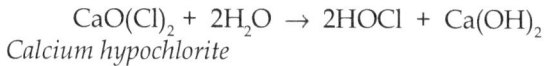

This hypochlorous acid is the actual disinfectant. It is a strong oxidant which is designated as antibacterial in nature. It is desirable to remove nitrogen or other contaminants during the secondary treatment before chlorination.

Disadvantages

More resistant types of organic molecules including some lipids and hydrocarbons are not oxidised completely but instead become, partially chlorinated. Chlorinated hydrocarbons tend to be toxic and are difficult to mineralise. Alternative means of disinfections are more expensive, hence chlorination remains the principle means of sewage disinfectant.

Review Questions

1. What is the aim of sewage purification
2. What is meant by stabilisation of sewage and how it is performed?
3. Describe the small-scale sewage treatment
4. Write notes on:
   a. Cesspool
   b. Septic tank
   c. Imhoff's tank
5. Explain in detail the large-scale sewage treatment.
6. Write notes on secondary treatment in sewered areas.
7. Write a note on the activated sludge process.
8. Write in detail about sludge digestion.
9. Write notes on tertiary treatment of sewage.

49

Applications of Waste Treatment

SACCHARIFICATION

Saccharification is the enzymatic conversion of starchy wastes into sugars thus taking care of a part of the waste recycling mechanisms. The following discussions highlight some examples of saccharification.

Wood Saccharification

From the earliest days of organic chemistry, scientists have been intrigued by the fact that most cellulosic wastes are two-thirds carbohydrate, which by appropriate treatment can be converted to sugars useful as food or for the preparation of chemicals.

The hydrolysis of cellulose to glucose and simple oligosaccharides will become a dominant economic factor. Cellulosic by-products will find increased utilisation as raw materials for the production of simple aliphatic chemicals through chemical engineering and biological processes. Looking far into the future we can see that cellulose, the most abundant of all photosynthetic products, will become one of great value. Cellulosic residues considered as a chemical raw material are a source of hexose and pentose sugars. The major portion, the hexoses, are equivalent to sugar from common sources. The pentoses, on the other hand, are unique, as they may be processed to furfural, a chemical which has not been produced from other sources. The pentosans in corncobs and bagasse are now the main source of furfural. The pentosan content of wood, however, is too low for an economic venture supported by this single product.

Purpose and scope Wood saccharification is but one aspect of a larger problem¾the chemical utilisation of cellulosic residues. Sugar is but one of many products obtainable from cellulosic residues, and wood is but one of a variety of potential starting materials.

Constituents of Wood

The amount and nature of the sugars obtainable from wood, and the processes required to effect the necessary hydrolysis are determined by the polysaccharides of the wood.

The main polysaccharide in all woody-plant materials is cellulose. This cellulose is chemically and physically similar to cotton in that it is fibrous, has a high resistance to alkali, and is hydrolysed only with much difficulty to yield glucose.

Hemicelluloses are present in wood and other cellulosic materials to an extent of roughly one-third of the total carbohydrate. The hemicellulose is amorphous and because of its lack of crystalline organisation, it hydrolyses much more easily than does cellulose.

SACCHARIFICATION PROCESSES

The conversion of cellulosic materials to sugar appears at first glance to be a simple hydrolytic cleavage of glycosidic bonds. In reality, cellulose is unique among the known polysaccharides in its extreme resistance to hydrolysis. The glycosidic bonds themselves are easily broken, but the crystalline organisation of the cellulose results in a low accessibility to dilute acid commonly used as a catalyst. As a consequence, the conditions of temperature and acid concentration required to accomplish the reaction in a reasonable time cause serious decomposition of the resulting sugars. Faced with these facts, only a few basic alternatives have presented themselves for practical hydrolysis.

1. A simple dilute-acid hydrolysis can be carried out without separation of product as it is formed.

2. A percolation process can be employed in which yields are raised by the expedient of continuously removing the product as it is formed.

3. A concentrated acid process can be used in which the crystalline organisation of the cellulose is destroyed, the carbohydrate solubilised, and finally hydrolysed completely with dilute acid.

All commercial processes fall into these three categories.

Simple Dilute-acid Hydrolysis of Wood

The single-stage batch process of wood hydrolysis was the first commercial method for making sugars from wood. The process has the advantage of great simplicity, and, with improvement, it might still be the method of choice in certain situations.

The wood is processed on a 1-hour cycle in four spherical digesters, each holding 4,700 pounds of dry wood. The sugar is extracted in a battery of eight 150-cubic-foot cells arranged for countercurrent extraction. About 96 percent of the sugar is extracted to give a solution containing about 12 percent total solids, nearly 9 percent reducing sugar, and roughly 6 percent fermentable sugar.

A process used in Sweden during World War II for the continuous hydrolysis of wastewood is described as follows. The wood chips are impregnated in a tower with sulphur dioxide gas. The bottom of the tower is connected to the feeder mechanism of a de-fibrator. The chips are heated in the pre-heater at 180° C for 2 to 3 minutes, ground in the de-fibrator, and the pulp pressed out continuously through an expansion valve. Seven such units operate in parallel. The pulp is washed in centrifuges and the solubles are re-acidified and hydrolysed further. The sugar solution at a concentration of 8 percent is fermented with the spent sulphite liquor of the plant. The residual lignocellulose is used as a fuel.

The following conclusions can be arrived at regarding the saccharification process involving wood:

1. A fairly good yield of sugar (48–50 percent);
2. A high concentration of sugar (10–12 percent);
3. All pentoses enriched in the first stage solution which facilitates the utilization of the pentoses: the sugar formed in the second stage consists entirely of glucose;
4. The total consumption of sulphuric acid is low, and amounts to 20–25 kilograms per ton of wood;
5. The consumption of steam without heat recovery amounts to 1.3 tons per ton of wood;
6. The digesters for carrying out the hydrolysis are of small size, because of the short time of hydrolysis. Fifty tons of wood a day may be

converted to sugar by using two digesters of 3 cubic metres each and two of 2 cubic metres each.

The hydrolysis is carried out under the following conditions:

1. The wood is used in the form of thin shavings, chips or sawdust.
2. The wood is impregnated with weak acid and freed from excess of liquor prior to hydrolysis.
3. The hydrolysis is carried out in an atmosphere of steam.

Details of the process

a. Impregnation with a liquid containing 0.5 percent sulphuric acid:

   Temperature: 190° C.

   Steam pressure: 12 kilograms/square centimetre.

   Time of hydrolysis: 3 minutes.

b. Washing out the sugar on a continuous counter current bed filter.

c. The residue from (a) is impregnated with a solution containing 0.75 percent sulphuric acid:

   Temperature: 215° C.

   Steam pressure: 20 kilograms/square centimetre.

   Time of hydrolysis: 3 minutes.

d. Washing out the sugar on a continuous counter current bed filter.

Percolation Processes

In saccharification by percolation processes, wood is put into an acid-resistant pressure vessel and hydrolysed by dilute acid injected into the top of the vessel and withdrawn through a filter at the bottom. In this way, sugar production and extraction go on simultaneously, and the sugar is separated and cooled as soon as possible to prevent decomposition.

The industrial development of the Scholler process using the percolation technique, has been described.

Typically, a Scholler plant has six or eight 50-cubic-metre digesters constructed of steel and lined with acid-resistant tile. The diameter of these digesters is 2.4 metres, and the overall height about 13 metres. The top of a digester, or percolator, has steam and vent lines and a line for the introduction of hot dilute acid. The bottom is equipped with a filter cone and a quick-opening discharge valve for removing the lignin residue. The digester is loaded with 9 to 10 metric tons of sawdust and chips to a density of 180 to 200 kilograms of dry wood substance per cubic metre. A charge of dilute acid is then injected at a temperature lower than that of the percolator contents, and the injected acid is heated by steam from the bottom until the desired temperature is reached. The solution is then pressed from the percolator by applying steam to the top of the charge. This operation is repeated for a total of up to 20 cycles, with 0.8 percent sulphuric acid at temperatures increasing to a maximum of 184° C. From 10 metric tons of wood approximately 120 metric tons of liquor with a concentration of 5 to 6 percent sugar is obtained.

Strong Acid Processes
for Wood Saccharification

These methods are characterised by the use of large amounts of concentrated acid which bring about extreme swelling of the cellulose. This serves to break the bonds which hold cellulose in the crystalline state and make it highly resistant to ordinary dilute acid hydrolysis. After extreme swelling and partial hydrolysis, the hydrolysis must be completed in dilute acid solution.

The very early work on cellulose saccharification made use of sulphuric acid. Wood is given a prehydrolysis with dilute acid to remove the hemicellulose. The dried residue is treated with sulphuric acid in an edge runner. This mixture is then diluted with the prehydrolysis liquor, and heated to complete the hydrolysis.

Advantages

1. High yield of alcohol.
2. Recovery of furfural and other eventual by-products such as acetic acid and ethereal oils.
3. Possibility of working with sawdust as well as with chips.

Disadvantages

1. High consumption of acid.
2. Necessity of solving the problem of placing large quantities of calcium sulphate.

Wood is hogged to chips not more than 1 cm in the long dimension, and conducted by a pneumatic conveyor to a rotary drier. Waste stack gas and the wood move parallel through the drier, and the moisture content is lowered to

6%. The wood is then loaded into 50-cubic-metre digesters lined with rubber- and acid resistant brick and extracted with 50% (by volume) hydrochloric acid. There are two parallel batteries of 14 extractors, half of each being extracted with concentrated acid, and the other half with water. The total time cycle per digester is 55 hours. As a result of the countercurrent extraction, with acid, a syrup is obtained consisting of water with 32% sugar, and 28% hydrochloric acid. This syrup goes to an evaporator system operating at 30 to 44 mm at 40° C where the sugar concentration is raised 60 to 63% and the acid concentration lowered to 2 to 5%. These evaporators have separate heaters in which syrup circulates inside the porcelain tubes. The head of the heater is of rubber-lined steel. Steam is then injected into the syrup to reduce the acid concentration. The carbohydrate in solution at this point consists primarily of oligosaccharides, and in order to convert it to monosaccharides it is 'inverted' by diluting and boiling. The residual acid is sufficient to catalyse the hydrolysis. The product is neutralised with lime and used for the production of yeast. The substances in solution consist of 70% glucose, 10% pentose, and 20% calcium chloride.

The recovered acid and the dilute acid washings pass to an acid recovery system where water is removed. This is accomplished by the addition of calcium chloride which increases the concentration of HCl in the gas phase. Water is continuously removed from the calcium chloride in the same apparatus. Conditions in this piece of equipment are extremely corrosive, and the operation is troublesome. The temperature is 145°C. In order to resist such conditions — high temperature, high acid concentration, and the presence of calcium chloride — the apparatus is lined with rubber and a double layer of brick. The heating tubes are of copper, plated with gold or platinum.

Clostridium thermoacellum which feeds on cellulose and *Thermoanerobacter ethanolicus* which feeds on starch saccharify the starch into fermentable sugar and may have the potential benefits in fuel alcohol production.

General Problems of Saccharification Processes

The problems associated with the development of an economic saccharification process for cellulosic materials are numerous and complex but they appear no more difficult than those faced by many of the present chemical industries during their development. The saccharification processes developed to date have met these problems in various ways but in none has complete cognizance been taken of all the difficulties, with the result that none have been economically successful without subsidy. The following discussion will cover the major points that should be borne in mind while weighing the merits of proposed processes and also point to the general direction in which research could be done.

The cellulosic raw materials may be broadly grouped into two classes: agricultural residues, which are harvested annually, and wood residues, which may be harvested continuously as desired. Both are available in continuous supply at low cost in the locality where they are grown. However, the annual crops usually are an expensive raw material when the cost of collection and storage is considered. Most of them also have a very low bulk density and large equipment and facilities are required. Annual crops have a further disadvantage in being subject to seasonal variations which require minor process variations. Wood is superior as a raw material in the general aspect to the annual crops but it too has disadvantages. It has a low bulk density, and the cost of handling and preparing wood for the saccharification step is high, even when using the best methods and equipment available. For instance, the handling of logs and conversion to chips in the Scholler process amounts to approximately one-half cent per pound of sugar produced.

The process should be matched carefully to the physical and chemical nature of the wood available. The strong acid process sets limitations on the raw material; sawdust often available at low cost (including handling) is unsuitable. Any process which requires bark-free chips incurs a large increase in raw material cost.

The bulk density of wood is an important factor in the plant cost. Saving could perhaps result from the development of a continuous process.

The only practical path from cellulose to sugar appears to be acid hydrolysis and this fact requires that all processes use acid at least throughout the hydrolysis step. This is a critical point for consideration, as the type of acid and its concentration has a very important effect on plant cost. In the strong hydrochloric acid process, the major portion of the plant must be acid-resistant and estimates of plant cost indicate a depreciation cost amounting to at least one-half cent per pound of sugar production on the basis of a large plant. This amounts to more than $200 of plant investment per annual ton of product, putting it in the class of expensive chemical plants. The plants using dilute acid require much less corrosion-resistant equipment and investment costs are somewhat less than half those quoted above.

Plant heat requirement depends largely on the end use of the hydrolysate. In the case where the product is crystalline glucose or molasses, the heat load is enormous. In the Bergius process, heat is required to recover the strong acid used for the primary hydrolysis, while in the dilute acid process the main consumption is in the evaporation of the dilute solutions to molasses. Methods for increasing the sugar concentration in the hydrolysates of the dilute-acid process have been employed and this gives an economic advantage to the dilute-acid process. In the event that the sugar is to be used in solution for the production of yeast, alcohol, or other product easily separable from dilute solution, the heat load is greatly decreased, giving dilute-acid processes a decided advantage over strong-acid processes.

Chemical costs for the dilute acid processes are small, amounting to less than one-fourth cent per pound of sugar. In the strong acid process, they become significant, as more than one-half cent per pound of sugar produced even after full advantage is taken of modern recovery methods. The chemical cost of the strong sulphuric acid process makes it unattractive except under certain circumstances where the acid might have further use.

Considering the high costs of handling the raw material, the chemical costs and the heating load, it is apparent that every effort should be made to obtain full utilisation of all the products available and these should be obtained in the highest possible quality compatible with cost. No commercial ventures have succeeded in such utilization. In no case has the lignin, which amounts to 20 to 30 percent of the dry wood substance, been used successfully other than as fuel. The hemicellulose fraction in some cases has been recognized as needing much milder treatment than the resistance cellulose, but no advantage has been taken of the fact that at least one unique chemical, furfural, may be obtained from this pentosan fraction.

In most cases the hexosan sugar product has been utilised as a crude molasses containing large quantities of unknown impurities. Some of these undoubtedly could be profitably removed as higher priced organic chemicals, resulting in a twofold gain. In the case of the strong acid process, a high-quality sugar fraction is obtained.

Recently a scheme has been proposed for manufacturing crystalline dextrose which would market approximately 55 % of the sugar product as food, the rest being marketed as molasses. The production of high-quality sugars from dilute acid wood hydrolysates is poorly understood and should be the subject of research. Another method is suggested for purifying the sugar by the addition of sodium chloride which forms a binary crystal with glucose. The yields of sugar from all processes are high. New research could profitably be done in the direction of increasing the yield of total products.

Waste disposal problems in an efficient saccharification process could be made small; molasses for stock feed and the concentration and burning of most of the other unwanted organics would solve most of the problems at moderate costs. One point worthy of mention here is the production of alcohol by the current Scholler process. The organic solubles and unfermented sugars in the still bottoms have an intolerably high biological oxygen demand. This is a case where the waste disposal problem would be nearly eliminated if the pentosan fraction were utilised.

GASIFICATION

Gasification is a process of thermal degradation of carbonaceous material under controlled amount of air or pure oxygen, and high temperature up to 1000°C. As a result of gasification, a large amount of gases is produced. Gasification of biomass is done in a gasifier designed in various ways. Success for gasification process is based on its designing. Therefore, the design of a gasifier is an important factor in controlling gas quality. Gas is used in a controlled manner for irrigation, pumping and electricity generation.

The solid waste stream can be divided into municipal solid waste (MSW) and industrial waste. MSW is a mixture of metals, glass, unmixed and mixed plastics, tyres, newspaper or paper and cellulosic waste (e.g. food, paper, yard waste). Source separation, collection and recycling mechanisms are currently available for metals, glass, unmixed plastics, and newspaper or paper with a moderate efficiency. The remainder of the MSW mixture goes to landfills.

Gasification, or the capture of energy from solid fuel and conversion into a clean combustible gas, has, until recently, proved elusive. Persistent challenges to gasification have included gasifier slag build-up, fouling, fuel inconsistency intolerance, the possibility of harmful by-products such as dioxins and heavy metals and prohibitive system costs. The extraction of maximum heat from a given fuel depends upon the efficiency of mixing the fuel with oxygen or air. This is perfectly achieved in the case of gaseous fuels. That is why conversion of solid waste into gaseous fuel is considered one of the best options. After pre-treatment it is fed into the main gasification chamber wherein biomass is converted into gas, which in turn, produces power after cooling and cleaning through gas engine connected to electric generator. A gasifier essentially carry out pyrolysis under limited air in the first stage followed by higher temperature reactions of the pyrolysis products to generate low molecular weight gases such as CO (carbon monoxide), CH_4, hydrogen, nitrogen, etc. The gas known as producer gas has a calorific value of 1000–1200 kcal/nm^3 which could be used in IC engines for direct power generation or in boilers for steam generation to produce power.

Process of Gasification

The MSW that is received is sorted to remove the non-combustible materials. The remaining organic fraction (ROF) is shredded to achieve the feedstock size of less than 5 cm. The moisture content of feed is maintained at less than 20% to maximise the heat recovery. Densification of ROF is not required, and this saves significant capital expenses and operating costs. The ROF is fed into the feed hopper with an agitator and hydraulically driven feed auger. The feedstock then passes to the main thermal reactor where at high temperature (900–1200°C) the ROF is converted into gas. Ash is removed from the base of reactor in a closed system. There are no fugitive airborne emissions from the ash systems. The gas flows from the top of the reactor vessel through pipes equipped with internal cleaners to mechanical gas cleaner vessel. Then the gas is passed through a series of mechanical cleaners to remove char, particulate matter, and any unreacted solids carried by the gas stream. The gas is then cooled through heat exchangers to the temperature as required by turbine, or boiler. A high voltage and low amperage electrostatic precipitator completes the gas cleaning and cooling process. A self-contained oil, tar or water separation system receives condensate from the electrostatic precipitator. Oils and tars are separated and re-injected into the reactor. The waste heat recovered from the reactor is used to preheat the feed-stock and reduce moisture to an acceptable limit (Solar Energy Research Institute, 1979). The quality of gas is regularly monitored through a gas meter. As per the requirement, the gas is fed to IC engines to produce power. The engine emissions are usually quite low. Another option is to use the gas in gas-fired boilers for steam generation, which in turn produces power through steam turbines.

Biomethanation (Anaerobic gasification)

Municipal solid waste is a heterogeneous waste and contains the following fractions:

1. *Putrescible fraction*   This is also called digestible fraction and contains biodegradable organic matter such as kitchen waste, vegetable market waste, paper, grass cutting, and yard trimmings. Putrescible fraction represents 40% of MSW in India.

2. *Combustible fraction*   Also known as refractory organics, these are either slowly digestible or indigestible organic matter such as wood, plastics, rubber, and other synthetics. They represent around 20% of MSW in India.

3. *Inert fraction*   They are typically non-digestible and non-combustibles such as stones, sand, glass, and metals. They represent 15% of the MSW in India.

4. Remaining 25% is the moisture content.

The putrescible fraction is ideally suited to produce biogas and the remaining slurry is a good fertiliser. There are different types of biogas technologies. MSW is subjected to mechanical segregation to obtain putrescible fraction, i.e. ROF. The digestible organic fraction thus obtained is kept as pulp in hydrolysis tanks for breaking them into smaller molecules. The hydrolysed pulp is then fed into anaerobic digestion tanks. Here it is digested anaerobically (in the absence of air), in the specially designed digesters. Under this active bacterial activity, the digested pulp produces the combustible gas CH_4, and inert gas, CO_2. The CH_4 gas is then used to produce power through a biogas engine connected to an electric generator. The remaining digestate (slurry) is a soil conditioner of good quality and free from pathogens. With the help of a solid or liquid separator, organic fertiliser is obtained and the treated water can be safely used for irrigation.

Depending upon the MSW quality, quantity and local environmental and climatic conditions, there are various commercially viable technologies available globally. In all the technologies the following steps are involved:

Sorting

The putrescible fraction is separated either manually or mechanically.

Particle Size Reduction

To provide maximum surface area to the bacteria, the particle size reduction is carried out by using screw cutting, milling, drumming, pulping, or shredding machines.

Digestion

The material is then fed into anaerobic digesters for gas generation.

Post-treatment

The slurry or digestate is matured for two to four weeks to make an agriculture or horticulture quality fertiliser or soil conditioner. The core of the whole technology is anaerobic digestion. There are many technologies for effective digestion which differ from each other depending upon their digestion parameters. They are briefly explained below:

Dry batch The MSW is fed batchwise in the digester after inoculating it with the slurry of the previous batch. The load is allowed to digest for 20–30 days till maximum gas is recovered. Commercially it is available as BOCELL process. It has the typical disadvantages of instability and material handling, pertaining to batch process.

Dry continuous DRANCO, VALORGA, and KOMPOGAS are the commercially available technologies in this category. They operate at the solid concentration of 20%–40% and they achieve high loading rates and minimise the requirement of water. Recycled effluent is mixed with fresh charge of MSW during semi-continuous feeding. They are ideally similar for thermophilic digestions (50–55 °C).

Wet continuous REFCOM technology uses the MSW solid concentration <10% using large amount of water. This is most ideal if mixed with sewage sludge, animal slurries, and industrial waste. To avail the large disposal of liquid, the effluent liquids are used to dilute the feed. This technology is usually not chosen if it is to be used only for MSW.

Wet multistage BTA and PAQUES designs are commercial technologies available in this category. MSW is converted into slurry with water and fermented in the first stage. The slurry with volatile acids is then converted to biogas in the second stage using high rate anaerobic digesters such as the anaerobic filter or UASB (upflow anaerobic sewage blanket) digesters. The major disadvantage is the complexity in design and operation. Highly skilled manpower is also required.

Sequencing batch This is a new concept known as SEBAC but yet to be fully commercialised. This is similar to the dry batch type technology except that leachate from the base of one digester is used to inoculate, and remove the volatile acid from other digester in the series. The digestion is usually allowed for 20–30 days in one digester and the digesters are filled every week in sequence so that continuous gas supply is available through the centralised gas collection system. High solid content over 30% is used in it.

The advantage of this design is that volatile acids are not allowed to accumulate and anaerobic bacteria keep on getting food continuously. The only disadvantage is the lack of continuous feed.

Status and cost Anaerobic digestion technology using mix-waste, bio-waste, and manure is working on a commercial basis in at least 15 countries, mostly in Europe, with capacity ranging from 4000 to 220 000 tonnes per year. Only a few of them work purely on MSW but the number of plants working solely on MSW is increasing, and many new ones are on the pipeline. There are some promoters such as ENKEM, CICON, GENL, and NSTLER who have offered to take up such plants using MSW on BOO basis in India. On an average they have estimated the total cost of the plant at 80–100 million rupees per 100 TPD capacity of the system. It has been observed that the cost of anaerobic digestion plants have reduced with time and is expected to be most competitive as compared to the other options. The most attractive aspect is environmental gain, i.e. no pollution and soil conditioning.

Merits and Demerits

An anaerobic digestion technology has many merits.

1. Useful products such as biogas and compost are obtained.
2. Drastic reduction of pathogens is achieved.
3. Needs comparatively much less land.
4. No release of greenhouse gases to environment.
5. No problem of odour.
6. Aesthetically good looking and hence no problem of real estate loss.
7. Comparatively more cost effective from the point of view of life cycle cost.

Besides having so many merits it suffers from some demerits.

1. It is a slow process and cannot accept shock loading.
2. Non-biodegradable organic fraction (i.e. refractory organics) cannot be digested and hence, needs to be disposed of without treatment or, additional technology (such as gasification) needs to be employed for the refractory organics.
3. The compost produced is not directly useful for soil, and needs special treatment to remove undesired materials such as glass, metals and pathogens. This adds to the cost of the plants.

Other Technologies

There are other technologies that have been reported to be tried on experimental levels but are yet to be commercially exploited. They are described below:

Pelletisation   Pelletisation is a process of producing fuel pellets from solid waste. The complete process involves drying, removal of non-combustibles, grinding, mixing, and production of pellets under high pressure. Usually, the conversion time is 25 minutes. The calorific value of raw garbage is around 1000 kcal/kg while the pellets also known as RDF have the calorific value around 4000 Kcal/kg.

About 15–20 tonnes of fuel pellets can be produced after treatment of 100 tonnes of raw garbage. These pellets could be used for heating in the boilers and the steam thus generated, in turn, is used to produce power. A power plant of 5 MW based on RDF will need 12 acres (4.85 hectares) of land and 600 TPD of raw garbage. It will cost 220–240 million rupees (1996 price). Two such plants are reported to be working in Bangalore and Mumbai.

Plasma arc (pyro-plasma process)   This system uses a heat source called a plasma arc flame. Two electrodes are precisely shaped and distanced. A highly ionised gas is passed between them and high voltage discharge occurs between the electrodes causing a hot plasma zone to be created. The plasma gas around the electrodes is of extremely high temperature ranging from 5600–30000°F (3093–16649 °C). At such a high temperature the molecules within that zone dissociate into their individual atoms. Thereafter, 'quenching' allows for the controlled cooling of the hot plasma gas. The reintegration process produces synthesis gas. Since the process occurs in the vacuum the intermediary products (pollution causing) NO_2 and SO_2 (oxides of sulphur) are not formed. It takes care of all organic matter whether biodegradable or not. The cost is expected to be Rs 40–50 million rupees. A fully commercial plant is yet to be installed. One such demonstration plant is expected to be installed in Taiwan.

Garret flash pyrolysis process   This low temperature pyrolysis process yields fuel oil. In this system, plant refuse is initially coarse shredded to less than 50 mm size, air-classified to separate organics, and dried in an air drier. The organic portion is then screened, passed through a hammer mill to reduce the particle size to less than 3 mm, and then pyrolysed in a reactor at atmospheric pressure. The heat exchange system allows pyrolytic conversion of the solid waste to a viscous at 500 °C.

Applications

With a wide range of acceptable feedstocks our waste gasification systems have a variety of potential applications, providing cost effective waste management solutions when you take into account disposal (tipping) fees and energy product sales.

- Municipal and industrial wastes are taken care of.

- All agricultural wastes can be made use of and well-disposed.

- Bio fuels can be produced from trees like poplar and willow, switchgrass, elephant grass, etc.

Vast global potential

- *Village electrification*   The World Bank estimates that today there are over 2 billion people, 400 million communities, without electric power. The market potential for our systems doubles when you include off-grid communities where electricity is derived from expensive (and highly polluting) fossil fuel sources. Conservatively a USD2 trillion potential.

- *Isolated grid 'mini utility'*   Applicable to industrialised and developing nations alike, waste gasification allows communities and industries to take charge of their wastes and create their own energy products. USD5 trillion conservative global potential.

- *Distributed power, grid connected, industrial waste utilisation*   Industries in developing nations as well as fully developed regions with modern infrastructure produce massive amounts of wastes. Most such wastes have a negative value (disposal costs) yet contain sufficient calorific value to be gasified and converted to valuable energy products. The advantages of gasifying these wastes is significant as industry can eliminate disposal fees, use energy internally and derive revenues from electricity sales to the grid. Landfill bound municipal solid waste in the G7 countries alone amounts to 16 billion tons per year, 80% of these wastes are usable in our gasification systems. These wastes, alone, would provide 700,000 MW of renewable electricity and define a potential market for 300,000 of our 150 TPD units or USD9 trillion global potential.

COMPOSTING

Composting is a method of combined disposal of refuse and night-soil or sludge. It is a process of nature whereby organic matter breaks down under bacterial action resulting in the formation of relatively stable humus-like material, called the compost which has considerable manurial value for the soil. The principal by-products are carbon dioxide, water and heat. The heat produced during composting, 60°C or higher, over a period of several days destroys eggs and larvae of flies, weed seeds and pathogenic agents. The end product, compost, contains few or no disease producing organisms, and is a good soil builder containing small amounts of the major plant nutrients such as nitrates and phosphates. The following methods of composting are now used: (1) Bangalore Method (Anaerobic method) (2) Mechanical composting (Aerobic method)

Bangalore Method (Hot Fermentation Process)

It is recommended as a satisfactory method of disposal of town wastes and nightsoil. Trenches are dug 90 cm (3 ft.) deep, 1.5 to 2.5 m broad and 4.5–5.1m long, depending upon the amount of refuse and night soil to be disposed of. Depths greater than 9 cm are not recommended because of slow decomposition. The pits should be located not less than 800m from city limits. The composting procedure is as follows:

First a layer of refuse about 15 cm thick is spread at the bottom of the trench. Over this, night-soil is added corresponding to a thickness of 5 cm. Then alternate layers of refuse and nightsoil are added in the proportion of 15 cm and 5 cm respectively, till the heap rises to 30 cm above the ground level. The top layer of refuse should be at least 25 cm thickness. Then the heap is covered with excavated earth. If properly laid, a man's legs will not sink when walking over the compost mass.

Within 7 days as a result of bacterial action considerable heat (over 60°C) is generated in the compost mass. This intense heat which persists over 2 or 3 weeks, serves to decompose the refuse and night-soil and to destroy all pathogenic and parasitic organisms. At the end of 4–6 months, decomposition is complete and the resulting manure is a well-decomposed, odourless, innocuous material of high manurial value ready for application to the land.

Mechanical Composting
(Indore Process)

Another method of composting known as 'mechanical composting' is becoming popular. In this, compost is literally manufactured on a large scale by processing raw materials and turning out a finished product. The refuse is first cleared of salvable materials such as rags, bones, metal, glass and items which are likely to interfere with the grinding operation. It is then pulverised in a pulverising equipment in order to reduce the size of particles to less than 2 inches. The pulverised refuse is then mixed with sewage, sludge or nightsoil in a rotating machine and incubated. The factors which are controlled in the operation are a certain carbon-nitrogen ratio, temperature, moisture, pH and aeration. The entire process of composting is complete in 4–6 weeks. This method of composting is in vogue in some of the developed countries.

Composting is a microbial process that converts putrefiable organic waste materials into a stable, sanitary, humus-like product that is reduced in bulk and can be used for soil improvement. Composting is accomplished in static piles, aerated piles, or continuous feed reactors. The static pile process is simple but relatively slow, typically requiring many months for stabilisation. Odour and insect problems can be controlled by covering the piles with a layer of soil, finished compost, or wood chips. Unless turned several times, the finished compost is rather uneven in quality. Under favourable conditions, self-heating in static piles typically raises the temperature inside a compost pile to 55–60°C or above in 2–3 days. After a few days at peak temperature, there is a gradual temperature decline.

Oxygen concentration in the compost is usually five times lower than in ambient air, even when the piles are mechanically turned. Some compost piles are often mechanically turned to maintain aerobic conditions. Turning of a compost pile may cause a secondary temperature rise brought about by the replenishment of the exhausted oxygen supply. Turning also helps to make the compost more uniform, because otherwise the thermophilic processes are restricted to the core of the compost pile. Following the thermophilic phase are several months of 'curing' at mesophilic temperatures. During this period, the thermophilic populations decline and are replaced by mesophiles that survived the thermophilic period. Because of the slowness of the composting process, large amounts of land are required, a disadvantage in densely populated urban areas.

The aerated pile process achieves substantially faster composting rates through improved aeration. The aeration is maintained by suction of air through perforated pipes buried inside the compost pile. This design achieves at least partial oxygenation of the pile, but temperature control is inadequate. Inside the pile, temperatures rise to self-limiting levels of 70–80°C. This can be improved by reversing the airflow from suction to injection. Thermostats placed inside the pile control blower operation, starting when the temperature exceeds 60°C. The injection of air not only oxygenates the pile but cools it sufficiently to avoid a self-limiting rise in the temperature. The heat generated by the biodegradation process is effectively used in evaporating water and results in a dryer and more stable compost. The aerated pile process goes to completion in about 3 weeks. Wood chips, if used as bulking agents, are removed from the final product by screening and are reused. The composting process could be hastened considerably by enriching the input airstream with pure oxygen. Although technically the concept of using pure oxygen seems attractive, it is highly doubtful that the returns would justify the sharply increased cost.

Composting can be accomplished more rapidly in a bioreactor. It requires about 20,000 cubic feet of air per ton or organic matter per day for efficient composting. This process forms a uniform and stable product, but it also requires a high initial investment. Composting in the reactor is accomplished in 2–4 days. A part or all of the reactor is maintained at thermophilic temperatures, using the heat produced in the composting process. After processing in the reactor, the product requires 'curing' for about a month prior to packaging and shipment.

Regardless of the process design, conducting the composting process in the thermophilic temperature range is desirable because it speeds the process and destroys pathogens that may be present in faecal matter and in sewage sludge. The aerobic oxidation reactions catalysed by microorganisms produce heat thus raising the temperature inside a compost pile to 76–78°C. Temperatures this high are actually inhibitory to biodegradation. Maximal thermophilic activity occurs between 52°C and 63°C. Aeration or turning may be adjusted to prevent excessive self-heating. Periodic water spraying can also reduce the temperature. Having sufficiently high temperatures is critical, however, for killing human pathogens because much of the material in compost piles, such as diapers, contains human faecal matter. Inadequate temperatures can lead to human health problems, especially if the piles are mechanically turned and particles become airborne.

The composting process is initiated by mesophilic heterotrophs. As the temperature rises, these are replaced by thermophilic forms. Thermophilic bacteria prominent in the composting process are *Bacillus stearothermophilus, Thermomonospora, Thermoactinomyces* and *Clostridium thermocellum.* Imporant fungi in the thermophilic composting process are *Geotrichum candidum, Aspergillus fumigatus, Mucor pusillus, Chaetomium thermophile, Thermoascus auranticus* and *Torula thermophila.*

Compost is a good soil conditioner and supplies some plant nutrients but cannot compete with synthetic fertilisers in agricultural production. If sewage sludge is a major component of the compost mixture, the finished compost may contain relatively high concentrations of potentially toxic heavy metals, such as cadmium and chromium. Compost finds unrestricted application in parks and gardens for ornamental plants and in land reclamation, such as for strip-mining reclamation and highway beautification projects.

Review Questions

1. What is saccharification?
2. Explain the process of saccharification.
3. What is gasification?
4. What is biomethanation?
5. Brief on the applications of biomethanation.
6. Describe the process of composting.

Glossary

Aerosol a fine suspension of particles or liquid droplets sprayed into air.

Algae a heterogeneous group of eukaryotic, photosynthetic organisms, unicellular or multicellular but lacking true tissue differentiation.

Allochthonous an organism or substance foreign to a given ecosystem.

Amensalism an interactive association between two populations that is detrimental to one and does not adversely affect the other.

Ammonification the release of ammonia from nitrogenous organic matter by microbial action.

Anaerobes organisms that grow in the absence of air or oxygen; organisms that do not use molecular oxygen in respiration.

Antagonism the inhibition, injury, or killing of one species of microorganism by another; an interpopulation relationship in which one population has a deleterious (negative) effect on another.

Aquatic growing, living in, or frequenting water; a habitat composed primarily of water.

Arbuscules specialised inclusions in root cortex cells in the vesicular-arbuscular type of mycorrhizal association.

Assimilation the incorporation of nutrients into the biomass of an organism.

Atmo-ecosphere that portion of the atmosphere in which living organisms are found and which is chemically transformed through the metabolism of organisms.

Atmosphere (atm) the whole mass of air surrounding the earth; a unit of pressure approximating 1×10^6 dynes/cm^2.

Autochthonous microorganisms and/or substances indigenous to a given ecosystem; the true inhabitants of an ecosystem; referring to the common micro biota of the body or soil microorganisms that tend to remain constant despite fluctuations in the quantity of fermentable organic matter.

Autotrophs organisms whose growth and reproduction are independent of external sources of organic compounds. the required cellular carbon being supplied by the reduction of CO_2 and the needed cellular energy being supplied by the conversion of light energy to ATP or the oxidation of inorganic compounds to provide the free energy for the formation of ATP.

B horizon the soil layer beneath the A horizon, consisting of weathered material and minerals leached from the overlying soil.

Bacteroids irregularly shaped (pleomorphic) forms that some bacteria can assume under certain conditions, e.g. morphological forms of rhizobia found in root nodules.

Barophiles organisms that grow best or

grow only under conditions of high pressure, e.g. in the ocean depths.

Barotolerant organisms that can grow under conditions of, high pressure but do not exhibit a preference for growth under such conditions.

Benthos the bottom region of aquatic habitats; collective term for the organisms living at the bottom of oceans and lakes.

Beta-oxidation (⍺-oxidation) metabolic pathway for the oxidation of fatty acids resulting in the formation of acetate and a new fatty acid that is two carbon atoms shorter than the parent fatty acid.

Biodegradable a substance that can be broken down to smaller molecules by microorganisms or their enzymes.

Biodisc system a secondary sewage treatment system employing a film of active microorganisms rotated on a disc through sewage.

Bio-geochemical cycling the biologically mediated transformations of elements that result in their global cycling, including transfer between the atmosphere, hydrosphere, and lithosphere.

Bio-leaching the use of microorganisms to transform elements so that the elements can be extracted from a material when water is filtered through it.

Biological control the deliberate use of one species of organism to control or eliminate populations of other organisms; used in the control of pest populations.

Biological oxygen demand (BOD) the amount of dissolved oxygen required by aerobic and facultative microorganisms to stabilise organic matter in sewage or water; also known as biochemical oxygen demand.

Biotic of or relating to living organisms, caused by living things.

Bloom a visible abundance of micro-organisms, generally referring to the excessive growth of algae or cyanobacteria at the surface of a body of water.

Breakpoint chlorination procedure for the removal and oxidation of, ammonia from sewage to molecular nitrogen by the addition of hypochlorous acid.

C horizon the soil layer beneath the B horizon consisting of the broken or partially decomposed underlying bedrock.

Carbon cycle the bio-geochemical cycling of carbon through oxidised and reduced forms, primarily between organic compounds and inorganic carbon dioxide.

Chemical oxygen demand (COD) the amount of oxygen required to oxidise completely the organic matter in a water sample.

Chemoautotrophs microorganisms that obtain energy from the oxidation of inorganic compounds and carbon from inorganic carbon dioxide; organisms that obtain energy through chemical oxidation and use inorganic compounds as electron donors.

Chemoorganotrophs organisms that obtain energy from the oxidation of organic compounds and cellular carbon from preformed organic compounds.

Chloramination the use of chloramines to disinfect water.

Chlorination the process of treating with chlorine, as in disinfecting drinking water or sewage.

Climax community the organisms present at the end point of an ecological succession series.

Coliforms gram-negative, lactose-fermenting, enteric rods, e.g. Escherichia coli.

Colonisation the establishment of a site of microbial reproduction on a material, animal, or person without necessarily resulting in tissue invasion or damage.

Colony the macroscopically visible growth of microorganisms on a solid culture medium.

Colony-forming units (CFUs) number of microbes that can replicate to form colonies, as determined by the number of colonies that develop.

Cometabolism the gratuitous metabolic transformation of a substance by a microorganism growing on another substrate; the cometabolised substance is not incorporated into an organism's biomass, and the organism does not derive energy from the transformation of that substance.

Commensalism an interactive association between two populations of different species living together in which one population benefits from the association, and the other is not affected.

Community highest biological unit in an ecological hierarchy composed of interacting populations.

Competition an interactive association between two species, both of which need some limited environmental factor for growth and thus grow at suboptimal rates because they must share the growth-limiting resource.

Composting the decomposition of organic matter in a heap by microorganisms; a method of solid waste disposal.

Corrosion the eating away of a metal resulting from changes in the oxidative state.

Cyanobacteria procaryotic, photosynthetic organisms contianing chorophyll a, capable of producing oxygen by splitting water; formerly known as blue-green algae.

Decomposers organisms, often bacteria or fungi, in a community that convert dead organic matter into inorganic nutrients.

Denitrification the formation of gaseous nitrogen or gaseous nitrogen oxides from nitrate or nitrite by microorganisms.

Detrial food chain a food chain based on the biomass of decomposers rather than on that of primary producers.

Detritus waste matter and biomass produced from decompositional processes.

Direct counting procedures methods for the enumeration of bacteria and other microbes that do not require the growth of cells in culture but rather rely upon direct observation or other detection methods by which the undivided microbial cells can be counted.

Direct viability count a direct microscopic assay that determines whether or not microorganisms are metabolically active, i.e. viable.

Diversity index a mathematical measure that describes the species richness and apportionment of species within the community.

Dormant an organism or a spore that exhibits minimal physical and chemical change over an extended period of time but remains alive.

Ecological balance the totality of the interactions of organisms within an ecosystem that describes the stable relationships among populations, and environmental quality.

Ecological niche the functional role of an organism within an ecosystem; the combined description of the physical habitat, functional role, and interaction of the microorganisms occurring at a given location.

Ecological succession a sequence in which one ecosystem is replaced by another within a habitat until an ecosystem that is best adapted is established.

Ecology the study of the interrelationships between organisms and their environments.

Ecosystem a functional self-supporting system that includes the organisms in a natural community and their environment.

Ectomycorrhizae a stable, mutually beneficial (symbiotic) assoication between a fungus and the root of a plant where the fungal hyphae occur outside the root and between the cortical cells of the root.

Effluent the liquid discharge from sewage treatment and industrial plants.

Endomycorrhizae mycorrhizal association in which there is fungal penetration of plant root cells.

Endosymbiotic a symbiotic (mutually dependent) association in which one organism penetrates and lives within the cells or tissues of another organism.

Enteric bacteria bacteria that live within the intestinal tract of mammals.

Enumeration determination of the number of microorganisms.

Epilimnion the warm layer of an aquatic environment above the thermocline.

Estuary a water passage where the ocean tide meets a river current; an arm of the sea at the lower end of a river.

Euphotic the top layer of water, through which sufficient light penetrates to support the growth of photosynthetic organisms.

Eutrophic containing high nutrient concentrations, such as a eutrophic lake with a high phosphate concentration that will support excessive algal blooms.

Eutrophication the enrichment of natural waters with inorganic materials, especially nitrogen and phosphorus compounds, that support the excessive growth of photosynthetic organisms.

Floc a mass of microorganisms cemented together in a slime produced by certain bacteria, usually found in waste treatment plants.

Food web an interrelationship among organisms in which energy is transferred from one organism to another; each organism consumes the preceding one and in turn is eaten by the next higher member in the sequence.

Freshwater habitats lakes, ponds, swamps, springs, streams and rivers.

Fungal gardens fungi grown in pure culture by insects.

Grazers organisms that prey upon primary producers; protozoan predators that consume bacteria indiscriminately; filter - feeding zooplankton.

Groundwater subsurface water in terrestrial environments.

Habitat a location where living organisms occur.

Halophiles organisms requiring NaCl for growth; extreme halophiles grow in concentrated brines.

Heterocysts cells that occur in the trichomes of some filamentous cyanobacteria that are the sites of nitrogen fixation.

Heterotrophs organisms requiring organic compounds for growth and reproduction; the organic compounds serve as sources of carbon and energy.

Host a cell or an organism that acts as the habitat for the growth of another organism; the cell or organism upon or in which parasitic organisms live.

Humic acids high-molecular-weight irregular organic polymers with acidic character; the portion of soil organic matter soluble in alkali but not in acid.

Humus the organic portion of the soil remaining after microbial decomposition.

Hydrosphere the aqueous envelope of earth, including bodies of water and aqueous vapour in the atmosphere.

Hypolimnion the deeper, colder layer of an aquatic environment; the water layer below the thermocline.

Indicator organism an organism used to identify a particular condition, such as *Escherichia coli* as an indicator of faecal contamination.

Indigenous   native to a particular habitat.

Leach   to wash or extract soluble constituents from insoluble materials.

Leguminous crop   plants belonging to the Leguminosae which have a seed pod divided into two parts or valves.

Lichens   a large group of composite organisms consisting of a fungus in symbiotic assoication with an alga or a cyanobacterium.

Limnetic zone   the portion of the water column in lakes excluding the littoral zone where primary productivity exceeds respiration.

Liquid wastes   waste material in liquid form which is the result of agricultural, industrial, and all other human activities.

Lithosphere   the solid part of earth.

Littoral   situated or growing on or near the shore; the region between the high and low tide marks.

Marine   of or relating to the oceans.

Mesophiles   organisms whose optimum growth is in the temperature range of 20-45°C.

Methanogens   methane-producing prokaryotes; a group of archaea capable of reducing carbon dioxide or low-molecular weight fatty acids to produce methane.

Microbial ecology   the field of study that examines the interactions of microorganisms with their biotic and abiotic surroundings.

Microbial mining   a mineral recovery method that uses bioleaching to recover metals from ores not suitable for direct smelting.

Microbial pesticides   preparations of pathogenic or predatory microorganisms that are antagonistic towards a particular pest population.

Microbiology   the study of microorganisms and their activities.

Microbiota   the totality of microorganisms associated with a given environment.

Microorganisms   microscopic organisms, including algae, bacteria, fungi, protozoa, and viruses.

Mineralisation   the microbial breakdown of organic materials into inorganic materials brought about mainly by microorganisms.

Most Probable Number (MPN)   a method for determination of viable organisms using statistical analyses and successive dilution of the sample to reach a point of extinction.

Mutualism   a stable condition in which two organisms of different species live in close physical association, each organism deriving some benefit from the association; symbiosis.

Mycelia   the interwoven mass of discrete fungal hyphae.

Mycobiont   the fungal partner in a lichen.

Mycorrhizae   a stable, symbiotic association between a fungus, and the root of a plant; the term also refers to the root-fungus structure itself.

Neuston   the layer of organisms growing at the interface between air and water.

Neutralism   the relationship between two different microbial populations characterised by the lack of any recognisable interaction.

Niche   the functional role of an organism within an ecosystem; the combined description of the physical habitat, functional role, and interactions of the microorganisms occurring at a given location.

Nitrification   the process in which ammonia is oxidised to nitrite and nitrite to nitrate; a process primarily carried out by the strictly aerobic, chemolithotrophic bacteria of the family Nitrobacteriaceae.

Nitrifying bacteria Nitrobacteriaceae; gram-negative, obligately aerobic, chemolithotrophic bacteria that oxidise ammonia to nitrate or nitrite to nitrate, and occur in aquatic environments and in soil.

Nitrite ammonification reduction of nitrite to ammonium ions by bacteria; does not remove nitrogen from the soil.

Nitrogen fixation the reduction of gaseous nitrogen to ammonia, carried out by certain prokaryotes.

Nitrogenase the enzyme that catalyses biological nitrogen fixation.

Nodules tumour-like growths formed by plants in response to infections with specific bacteria within which the infecting bacteria fix atmospheric nitrogen; a rounded, irregularly shaped mineral mass.

O horizon the organic layer of soil, consisting of humic substances.

Oligotrophic lakes and other bodies of water that are poor in those nutrients that support the growth of aerobic, photosynthetic organisms; microorganisms that grow at very low nutrient concentrations.

Oxidation pond a method of aerobic waste disposal employing biodegradation by aerobic and facultative microorganisms growing in a standing water body.

Ozonation the killing of microorganisms by exposure to ozone.

Paralytic shellfish poisoning caused by toxins produced by the dinoflagellate *Gonyaulax*, which concentrates in shellfish such as oysters and clams.

Parasites organisms that live on or in the tissues of another living organism, the host, from which they derive their nutrients.

Parasitism an interactive relationship between two organisms or populations in which one is harmed and the other benefits; generally, the population that benefits(the parasite) is smaller than the population that is harmed.

Pathogens organisms capable of causing disease in animals, plants, or microorganisms.

Pelagic zone the portion of the marine environment beyond the edge of the continental shelf comprising the entire water column but excluding the sea floor.

Pest a population that is an annoyance to economic health, or aesthetics.

Pesticides substances destructive to pests, especially insects.

pH the symbol used to express the hydrogen ion concentration, signifying the logarithm to the base 10 of the reciprocal of the hydrogen ion concentration.

Photoautotrophs organisms whose source of energy is light and whose source of carbon is carbon dioxide, and includes plants, algae, and some prokaryotes.

Photolysis liberation of oxygen by splitting of water during photosynthesis.

Photosynthesis the process in which radiant (light) energy is absorbed by specialised pigments of a cell and is subsequently converted to chemical energy; the ATP formed in the light reactions is used to drive the fixation of carbon dioxide, with the production of organic matter.

Phototrophs organisms whose sole or principal primary source of energy is light; organisms capable of photophosphorylation.

Phycobiont the algal partner of a lichen.

Phytoplankton passively floating or weakly motile photosynthetic aquatic organisms, primarily cyanobacteria and algae.

Phytoplankton food chain a food chain in aquatic habitats based on the conusmption of primary producers.

Plankton collectively, all microorganisms and invertebrates that passively drift in lakes and oceans.

Plate counting method of estimating number of microorganisms by diluting samples. Culturing on solid media, and counting the colonies that develop to estimate the number of viable microoganisms in the sample.

Pollutants materials that contaminate air, soil, or water substances and foul water or soil, thereby reducing their purity and usefulness. They are often harmful..

Predation a mode of life in which food is obtained primarily by killing and consuming animals, an interaction between organisms in which one benefits and one is harmed, based on the ingestion of the smaller organism, the prey, by the larger organism, the predator.

Predators organisms that practice predation.

Pre-emptive colonisation alteration of environmental conditions by pioneer organisms in a way that discourages further succession.

Prey an animal taken as food by a predator.

Primary producers organisms capable of converting carbon dioxide to organic carbon, including photoautrotophs and chemoautotrophs.

Primary sewage treatment the removal of suspended solids from sewage by physical settling in tanks or basins.

Profundal zone in lakes, the portion of the water column where respiration exceeds primary productivity.

Pro-cooperation synergism a non-obligator relationship between two microbial populations in which both populations benefit.

Protozoa diverse eukaryotic, typically unicellular, nonphotosynthetic microorganisms generally lacking a rigid cell wall.

Recalcitrant a chemical that is totally resistant to microbial attack.

Red tides aquatic phenomenon caused by toxic blooms of *Gonyaulax* and other dinoflagellates that colour the water and kill invertebrate organisms; the toxins concentrate in the tissues of filter-feeding molluscs, causing food poisoning.

Rhizosphere an ecological niche that comprises the surfaces of plant roots and the region of the surrounding soil in which the microbial populations are affected by the presence of the roots.

Rhizosphere effect evidence of the direct influence of plant roots on bacteria, demonstrated by the fact that microbial populations usually are higher within the rhizosphere (the region directly influenced by plant roots) than in root-free soil.

Rotating biological contactor see biodisc system.

Secondary sewage treatment the treatment of the liquid portion of sewage containing dissolved organic matter, using microorganisms, to degrade the organic matter that is mineralised or converted to removable solids.

Septic tank a simple anaerobic treatment system for waste water where residual solids settle to the bottom of the tank and the clarified effluent is distributed over a leaching field.

Sewage the refuse liquids or waste matter carried by sewers.

Sewage treatment to reduce the biological oxygen demand of sewage and to inactivate the pathogenic microorganisms present in it.

Sludge the solid portion of sewage.

Soil horizon a layer of soil distinguished from layers above and below by characteristic physical and chemical properties.

Solid waste refuse waste material composed of both inert materials↑ glass, plastic, and metal↑ and decomposable organic wastes, including paper and kitchen scraps.

Sparge to pass a gas through a solution.

Succession the replacement of populations by other population better adapted to fill the ecological niche.

Sulphur cycle bio-geochemical cycle mediated by microorganisms that change the oxidation state of sulphur within various compounds.

Symbiosis an obligatory interactive association between members of two populations, producing a stable condition in which the two organisms live together in close physical proximity to their mutual advantage.

Symbiotic nitrogen fixation fixing of atmospheric nitrogen by bacteria living in mutually dependent associations with plants.

Synergism in antibiotic action, when two or more antibiotics are acting together, the production of inhibitory effects on a given organism that are greater than the additive effects of those antibiotics acting independently; an interactive but nonobligatory association between two populations in which each population benefits.

Syntrophism the phenomenon that occurs when two organisms mutually complement each other in terms of nutritional factors or catabolic enzymes related to substrate utilisation; also termed cross-feeding.

Tertiary sewage treatment a sewage treatment process that follows a secondary process, aimed at removing non-biodegradable organic pollutants and mineral nutrients.

Thermocline zone of water characterized by a rapid decrease in temperature, with little mixing of water across it.

Trickling filter system a simple, film-flow aerobic sewage treatment system; the sewage is distributed over a porous bed coated with bacterial growth that mineralises the dissolved organic nutrients.

Trophic level the position of an organism or population within a food web; primary producer, grazer, predator, etc.

Ultraviolet light (UV) short wavelength electromagnetic radiation in the range 100–400 nm.

Vesicular-arbuscular mycorrhizae a common type of mycorrhizae characterised by the formation of vesicles and arbuscules.

Viable plate count method for the enumeration of bacteria whereby serial dilutions of a suspension of bacteria are plated onto a suitable solid growth medium, the plates are incubated, and the number of colony-forming units is counted.

Virus a noncellular entity that consists minimally of protein and nucleic acid and that can replicate only after entry into specific types of living cells; it has no intrinsic metabolism, and its replication is dependent on the direction of cellular metabolism by the viral genome; within the host cell, viral components are synthesised separately and are assembled intracellularly to form mature, infectious viruses.

Volatile organic chemical (VOC) organic compound that vaporises into the atmosphere.

Xenobiotic a synthetic product not formed by natural biosynthetic processes; a foreign substance or poison.

Yeasts a category of fungi defined in terms of morphological and physiological criteria; typically, unicellular, saprophytic organisms that characteristically ferment range of carbohydrates and in which asexual reproduction occurs by budding.

Zymogenous term used to describe opportunistic soil microorganisms that grow rapidly on exogenous substrates.

References

Ahmadjian. V. 1963. "The Fungi of Lichens." *Scientific American.* 208: 122-132.

Alexander, M. 1971. *Microbial Ecology.* Wiley, New York.

Alexander, M. 1977. *Introduction to Soil Microbiology.* 2 e. Wiley, New York.

Alexander, M. 1977. *Introduction to Soil Microbiology.* New York:

Alexander, M. 1977. *Introduction to Soil Microbiology.* Wiley, New York. pp 148-202.

Alexander, M. 1977. *Introduction to Soil Microbiology.* John Wiley and Sons, New York.

Andersen, A.A. 1958. "A new sampler for the collection, sizing, and enumeration of the viable airborne bacteria." *Journal of Bacteriology.* 76: 471-484.

Atlas, R. M. 1992. "Detection and enumeration of microorganisms based upon phenotype". In M.A. Levin, Seidler. R. J.(eds.). *Microbial Ecology, Principles, Methods, and Applications.* McGraw-Hill, New York. pp. 29-43.

Bagyaraj, D. J. and Varma, A. 1995. "Interaction between arbuscular mycorrhizal fungi and plants." *Advances in Micobial ecology.* 14: 119-142.

Barnett, H. L. 1963. "The nature of mycoparasitism by fungi." *Annual Review of Microbiology.* 17: 1-14.

Bartholomew, G. W. and Alexander, M. 1981. "Soil as a sink for atmospheric carbon monoxide." *Science.* 43: 1389-1391.

Baylor, E.R., Peters,V. and Baylor, M.B. 1977. "Water-to-air transfer of virus." *Science,* 197: 763-764.

Belser, L.W. 1979. "Population ecology of nitrifying bacteria." *Annual Review of Microbiology.* 33: 309-334.

Bolin, B. 1970. "The carbon cycle." *Scientific American.* 223 (3): 125-132.

Bonde, G. J. 1977. Bacterial indicators of water pollution. *Advances in Aquatic Microbiology.* 1: 273-364.

Breznak, J. A. and Pakratz, H. S. 1977. "*In situ* morphology of the gut microbiota of wood-eating termites." *Applied and Environmental Microbiology.* 33: 406-426.

Brierley, C. L. 1982. "Microbiological mining." *Scientific American,* 247 (2): 44 – 53.

Brierley, C.L. 1978. "Bacterial leaching CRC." *Critical Reviews in Microbiology,* 6: 207 – 262.

Brill, W. J. 1979. "Nitrogen fixation: Basic to applied." *American Scientist.* 67: 458 – 466.

Brock, T.D. 1978. *Thermophilic Microorganisms and Life at High Temperatures.*

Brown, M. E. 1974. "Seed Bacterisation." *Annual Review of Phytopathology.* 12: 311 – 331.

Bungay, H. R. and Bungay. M. L. 1968. "Microbial interactions in continuous culture." *Advances In Applied Microbiology.* 1: 269 – 290.

Buol, S. W. 1973. *Soil Genesis and Classification.* Oxford and IBH Publishing Co.

Cairns, J. (ed.). 1970. *The Structure and Function of Freshwater Microbial Communities.* American Microscopical Society Symposium, Research Division Monograph 3. Virginia Polytech Institute and State University, Blacksburg.

Caldwell, D.E. 1977. "The planktonic microflora of lakes." *Critical Reviews in Microbiology.* 5: 305 – 370.

Centiganto, Y. M. and Silver. W. S. 1964. "Leaf nodule symbiosis. I Endophyte of Psychotria bacteriophila." *Journal of Bacteriology.* 88: 776 – 781.

Cheng, L. 1975. Marine pleuston: "Animals at the sea-air interface." *Annual Reviews in Oceanography and Marine Biology.* 131: 181 – 212.

Churchill, B. W. 1982. "Mass production of microorganisms for biological control." In *Biological control of weeds with plant pathogen.* R. Charudattan and H. Walker (eds.). 139 – 56. John Wiley and Sons, New York.

Cole, G. A. 1983. "Textbook of Limnology." Mosby, St. Louis, MO.

Daji, J. A. 1968. "*A Textbook of Soil Science.*" Media Promoters and Publishing, Bombay.

David, P. M. 1965. "The surface fauna of the ocean." *Endeavor.* 24: 95 – 100.

David, P.M. 1965. "The surface fauna of the ocean." *Endeavor.* 24: 95 – 100

Deevey, E. S. 1970. "Mineral cycles." *Scientific American.* 223 (3): 149 – 158.

Devau, K. E. 1956. "Mutual relationships in fungi." *Annual Reviews of Microbiology.* 10: 115 – 140.

Diaz, L. F. and Goleuke, C. G. 1993. *Composting and Recycling*: Municipal Solid Waste. CRC Press, Boca Raton, FL.

Fenchel, T. 1969. *The Ecology of Marine Microbenthos.* Pat IV. Ophelia 6: 1 – 182.

Finstein, M. S. and Morris, M. L. 1975. "Microbiology of municipal solid waste composting." *Advances in Applied Microbiology.* 19: 113 – 151.

Fletcher, M. 1979. "The attachment of bacteria to surfaces in aquatic environments." In D.C. Ellwood, J. Melling and P. Rutter (eds.). *Adhesion of Microorgnaisms to Surfaces.* Academic Press, London. pp. 87‑108.

Ford, T. E. (ed.). 1993. *Aquatic Microbiology:* "An Ecological Approach." Blackwell, Oxford, England.

Foth, H. D. 1984. *Fundamentals of Soil Science.* 7e. John Wiley and Sons, New York.

Gallon, J. R. and Chaplin, A. E. 1987. *An introduction to Nitrogen Fixation.*

Gilman, J.C. 1945. *A Manual of Soil Fungi.* Collegiate Press, Ames.

Gray, T. R. G. and Parkinson, D. 1968. *The ecology of Soil Bacteria.* University of Toronto Press, Toronto.

Havelka, U. D. and Hardy, R. W. F. 1982. "Biological Nitrogen fixation." In *Nitrogen in agricultural soils,* F. J. Stevenson (eds.). 365‑422. American Society of Agronomy, Madison, WI.

Hungate, R. E. 1978. "The rumen microbial ecosystem." *Annual Reviews of Microbiology.* 29: 39‑66.

Hynes, H. B. N. 1970. *Ecology of Running Waters.* Liverpool University Press, Liverpool, England.

John Everett Park. 1995. *Park's Textbook of Preventive and social Medicine.* 14e edition. John Wiley and Sons, New York.

Kellog, W. W. and Cadle, R. D. 1972. "The sulfur cycle." *Science.* 175: 587‑596.

Kolay, A. K. 1993. *Basic Concepts of Soil Science.* Wiley Eastern Limited.

Kuenen, J. G. and Gemerden, H. V. 1985. "Microbial interactions among aerobic and anaerobic sulfur-oxidizing bacteria." *Advances in Microbial Ecology.* 8: 1‑59.

Lewin, R. 1984. "How microorganisms transport iron." *Science.* 225: 401‑402.

Mack, W. N. and Ackerson, O. 1975. "Microbial film development in a trickling filter." *Microbial Ecology.* 2: 215‑226.

Mehtar, S. 1992. *Hospital Infection Control – Setting up a Cost Effective Programme.* Oxford University Press, Oxford.

Natesh, S., Chopra, V. L. and Ramachandran, S. 1987. *Biotechnology in Agriculture.* Oxford and IBH Publ. Co., New Delhi.

Nedwell, D. B. 1984. "The input and mineralization of organic carbon in anaerobic aquatic sediments." *Advances in Microbial Ecology.* 7: 93‑131.

Neilands, J. B. 1981. "Microbial iron compounds." *Annual Review of Biochemistry.* 50: 715-731. Springer-Verlag, New York.

Odum, E. P. 1971. *Fundamentals of Ecology.* Oxford. Saunders, Philadelphia.

Olson, B. H. and Nagy, L. A. 1984. "Microbiology of potable water." *Advances In Applied Microbiology.* 30: 73–132.

Rangaswami, G. 1988. "Soil plant microbe interrelationship." *Indian Phytopathology.* 41: 165–172.

Rheinheimer, G. 1991. "Aquatic Microbiology." Wiley, New York.

Schardl, J. J. and Hall, D. O. 1988. "The *Azolla-Anabaena* association: Historical perspective, symbiosis and energy metabolism". *Botanical Review.* 54: 353–386.

Singh, R. N. 1961. *The Role of blue-green algae in nitrogen economy of Indian agriculture.* ICAR, New Delhi.

Subrahmanyam, N. S. and Sambamurty, A.V.S.S. 2000. *Ecology.* Narosa

Taylor, W. R. 1942. "Bacteriology of freshwater." *Hygiene.* 42: 284–296.

Weber, N. A. 1966. "Fungus-growing ants." *Science.* 153: 587–604.

Wilson, G. and Pike, E. B. 1990. "The bacteriology of water." In Parker M. T and Collier L. H. (eds). *Topley and Wilson's Principles of Bacteriology.* pp. 243–264. Edward Arnold, London.

Wolf, F.T. 1943. "The microbiology of the upper air." *Bulletin of the Torrey Botanical Club.* 70: 1–14.

Index